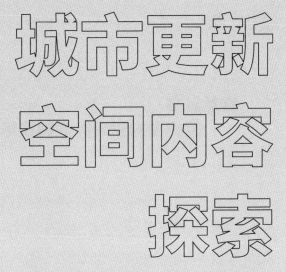

城市更新空间内容探索

Urban Renewal Content Space Exploring

毛大庆　贾晓萌　王　翔　编著

中国建筑工业出版社

序　言

存量城市
如何重构空间价值的续写方式

空间与时间，是文明诞生之日起人类恒久的主题。

如何将时间嵌入空间，让人们在空间内更好地感知时间、利用时间、消费时间甚至优雅地虚耗时间，是我们在城市更新过程中必须认真思考的问题。

这是一本关于城市更新的书，也是一本关于空间与内容的书。

本书中展示了近几年共享际和优客工场这两家致力于城市更新的创新型企业所做的各种有益尝试。从中，我们对空间有了各种精彩的解读与重构。

城市更新的主旋律下，一面是共享际，一面是优客工场。我们需要以更加开放的眼光来看待我们的生活与城市，空间价值的重构也成了这一代建筑师的使命，这里存在两个问题：一是对既有空间的认知体系；二是对空间未来属性的重新定义。

这里不仅仅是时间分配的大原则，还需要我们重新解读"城市更新"；不仅仅是空间与时间层面的共享，也是针对资源的优化与配置，还需要我们通过平台进行有效的资源整合，重新共建；从而使原来低效、低收益的资源产生出更大的效益，这是我们对城市更新的理解。

1．在空间与时间的交互结构中阐明设计理念，构建生活场景，满足一切与生活内容相关的需求

空间与时间，是文明诞生之日起人类恒久的主题。

如何将时间嵌入空间，让人们在空间内更好地感知时间、利用时间、消费时间甚至优雅地虚耗时间，是我们在城市更新过程中必须认真思考的问题。

要想回答这个问题，设计与设计感是我们不可能绕过的一个重要课题。

设计体现了人与物的关系，设计是人类本能的体现，是人类审美意识的驱动，是人类进步与科技发展的产物，是人类生活质量的保证。

唯其如此，商业空间的设计，也就成了城市更新过程中一个极其重要的抓手。设计不仅是一种语言，还是一种动作，更是一种体验，借由设计，建筑物的定位与主题拥有了生命力，并时刻向来来往往的人们展示着这座城市的品位。

无论是优客工场所聚焦的商业办公空间，还是共享际所专注的生活方式体验的复合空间，究其本质，都具备城市空间交互的属性。

城市空间的概念源于人类特有的人文环境，它不仅要满足人的个人需求，还应满足人与人的交往及其对环境的各种要求。

城市空间所面临的服务对象涉及不同文化素养层次、不同职业、不同种族，是社会化的行为场所。

尤其是在城市更新过程中，在地居民与旅居者之间的包容与冲突，必然会通过空间设计得以推动或化解，因为空间最重要的功能就是社交活动，在社交过程中才有可能产生消费场景，从而带来空间价值的提升。

这些年，共享际与优客工场参与了大量的城市空间改造项目，从设计师的角度讲，改造空间的设计难度，要比新建空间大很多。这个道理不言自明，至少从主观上讲，在白纸上作画，总比在成画上添墨绘彩，可施展的空间要大很多。

改造一个项目前，我们必须先搞清楚三件事：
第一，这个空间的未来功能属性是什么？将承载什么内容？
第二，成本核算总是最难的，因为绝大多数改造项目的预算很低，我们要在有限的预算内实现最好的效果。
第三，改造项目的现状往往与图纸大相径庭，必须要实地勘测，深入工地去理解空间的真实结构。

优客工场的很多共享办公空间都是从老旧厂房、超市等废弃商业场所改造而来的，办公空间的设计与其他类型商业空间的理念天差地别，我们的设计师不但要考虑空间的利用率和结构的合理性，还要考虑功能区域的布局，以会议功能为例，设计师不但将共享的理念植入会议空间，让这一功能更加高效，还要考虑各种规格的配置与场景；针对入驻比例较高的中小微、初创企业，我们还在空间内设置了大量的小型交流单元，而大型空间也尽量以移动墙体等建筑材料实现其空间的灵活分割。

共享际的众多项目都藏身于北京内城的胡同中。胡同建筑的设计改造又是另一门学问，一方面要与胡同内的老北京在地居民达成足够的谅解与融合，另一方面建筑改造后的风格又不能与在地文化产生过于明显的冲突。在设计方面，比如胡同内建筑需要更多考虑自然光的利用与加持，来呈现一个阳光普照的生活空间，以及活动空间等特色场景设计，在凝集社群文化的同时，更加多样化地探索在地文化，这一切都让项目变得如此与众不同。

这就是设计的力量。

共享际与优客工场，从人性需求的设计出发，在城市更新的过程中，展现了更多的可能性。

2. "共享际"与"优客工场"，各有侧重点，但终将回归再利用与再塑造的主题

六年多前，我发起创立了两家公司，"优客工场"和"共享际"。优客工场目的简单一些，它是一种城市工作方式的引领；共享际，偏向生活方式，所做更多的是一种价值引领。

然而，在本书中更想表达的是：再利用、再塑造的主要推动力量不再是建造，而在保留为前提的基础上根据场景去赋予空间新的呈现形式，这可能是空间内容跟功能的一种叠合方式，或者说是空间内容加功能之后形成的一种生活方式，我们需要来破解一些题目。

其实，现在很多定位"城市中心"的运营方都在破题，出给我们的题目是在这样的地理位置、这样的城市生活场景下，有一个空置或闲置的低效资产，需要我们去破题。

共享际不同于优客工场只针对工作场景，相对单一的内容，我们将问题延伸到生活场景时，共享际对城市空间内容运营应运而生，诸如餐饮、游戏、戏剧、阅读、体育、乡村等方面有更为广大的课题，远远不是办公能够填充的。这里面有很多社群的反向定义，比如，我们发现戏剧爱好者的数量日益增多，他们对戏剧消费的需求构成了很多对空间内容的填充；再比如，很多人对体育、音乐类的需求增多，又会反向填充对空间的这种需要。

今天产生了许多比较有趣的现象，像林下经济、林草地、废弃用地这些可能连房子都不需要的场景，催生了营地产业。包括一些儿童的自然科普教育、自然地理课堂等，实际上它已经超越了建筑物，是很多实际的生活需求所构成的、搭建的消费场景。

因为需求产生了很多空间内容，所以我们要有这样一个载体，把价值的构成植入进去，输出为各个不同的城市场景，称之为"给建筑物赋能的能力"，无论是城市管理者还是商业运营商都非常需要这种载体。

回归到空间的本质，也就是在新城市空间、新乡村空间或者是城乡接合部的状态下，我们能输出的、给予的内容，到底有哪些？

3．从城市走向乡村，我们更应该关注的是人，而不是建筑

城市之外，也是我们关注和向往的地方，在城市化蓬勃发展的今天，乡村的产业和空间更新也需要我们从人的角度出发，创造适合今天城市人和乡村人共同生活的"远方"。

前不久，我带领共享际的设计团队参加了一个民宿设计评比活动，发现一些设计师会站在运营的视角探讨民宿的设计与未来发展的规划，会考虑与在地村民的融合共生，会思考如何通过民宿为农村提供更多的经济支撑，包括就业、农副产品销售等。假设把纯建筑感的东西放在农村，是存在一些隐患和风险的。农民基于他们已有的认知，可能无法理解这种风格，甚至与在地风格发生巨大的认知冲突，从而产生负面情绪与负面效果。

乡村民宿的理想意义应当是缩小城乡差别，提升农村的美誉度和艺术鉴赏力，同时我们更多的是在强调宿集而不是民宿，因为它应该是一个集群化的东西，是一个内容集合体，而不再是个案。每一个民宿可以由不同的主理人决定其发展方向，但最终应该是整体IP和平台经营，以实现整体的价值输出，这样既可以赋能于这一批主理人，同时又能够真正地向乡村赋能，这才是宿集发展之路。

乡村建设与更新面临着比城市中的改造更新更大的难题与挑战，例如庞大的在地人群与交通配套设施等都无法与城市相比，因此乡村的建设与更新就需要走出一条全新的经营之路，创新农业，专注体验，设计赋能，让逆城市化的人潮回到一个田园版的乡村。

4．城市更新背后的深层逻辑来源于对旧资源的再次共享

当社会经济高速发展时，大量的存量资产出现，这其中不仅有房子，还有各种低效内容和低效产品。再次共享可以很好地消化与利用这些资产，通过平台的能力进行整合并激发新的活力。

我认识一位北京的老厂长，是做资源再利用的。他把各种无机的垃圾搅碎，通过各种环保材料再融合，从而产生出各种新的建筑材料，最后都可以放在各种各样的场景下使用。老厂长有一句话令我记忆犹新：在我的眼里，天下没有垃圾。

老厂长带给我的启示，让我可以更加深入地投入到优客工场和共享际的事业当中——其实我们真的不需要再去做更多增量的资源。增量资源从何而来?是由于人们的惰性，旧的不用，新的源源不断，这就是所谓的增长。在城市高速发展的过程中，我们有必要停下来，认真反思"更新"的定义，不断拷问自己：

为什么要更新?

为谁而更新?

拿什么去更新?

如何更新?

更新到什么样才算成功?

这两年，很多人言必称城市更新，可又有几个人能对如上问题做出令人满意的回答呢?

写在最后：这个时代，最缺的是慢下来的好物

算起来，我专注于优客工场与共享际的事业，持续的时间并不算长，不过六年有余，但于我而言，其意义却更加深远。

关于优客工场和共享际目前的架构设置，虽然两个品牌都主打共享，但却是两个完全独立的团队，并没有太直接的关联，它们是两个不同赛道上的业务，但核心都是城市更新。

共享际偏重于价值输出，优客工场是已经形成固定状态的操作流程。共享际的IP特别多，所以预计未来输出的子品牌应该会很多，而且会越来越多。这样不管放到乡村还是城市，它的状态相比之下会更灵活。

无论优客工场还是共享际，我们要探讨、反思的都是如何在存量中寻找增量的命题，这个命题的原点是我们对于国家、城市、乡村的初心。

以前是从甲方到乙方，再从乙方返回到甲方的销售思维，现在是先有共同体的价值思考状态，然后再形成空间状态。

传统开发模式的特点是快节奏、快速转化，而共享际则是需要我们沉心静气地投入其中，用时间的力量去打磨与渗透。

这个时代从不缺快节奏的东西，恰恰，一些慢下来的好物，才更显出它的价值。

于是，我的身份是一个判断者和价值的挖掘者。

关于一座城市中存量空间价值续写的命题，仍在继续。

毛大庆

优客工场/共享际创始人、董事长

建筑师

贾晓萌

共享际执行CEO/联合创始人

王　翔

共享际设计总监/首席建筑师

推荐序一

城市不是明信片
是生活的载体

在我看来，优客工场与共享际所参与的城市更新项目，是城市开发从"投资"到"使用"空间的转变，一个我们虽然还未介入但迟早都要深入了解的新阶段。

优客工场和共享际是两家以不同视角切入城市更新的创新企业。在这本以城市与空间为主题的精美图册中，呈现的就是几年来借由它们所缔造的特色鲜明的城市空间，这些空间除了带给我非常强烈的视觉冲击力之外，我还能感受到一种有关"年轻"的能量。

年轻人是起始也是目标，更是能量的源泉。

在中国社会和城市的飞速发展过程中，认清当下、预判未来，绝对需要一种长期学习与独立思考的洞察力。房地产发展的过量问题，其实在很多年前已有预兆。今天看来，存量建筑的确是一个在不断放大的紧迫问题。

除了存量建筑之外，我们知道，城市中还有很多老房子，它们与开发商无关，却在快速发展的时代中几乎被遗忘。如何发掘与培育这些老房子的活力，是一门非常讲究的学问。

我本人常年在北京的方家胡同工作，对北京传统文化和城市古建筑有着较高的关注度。我不想具体评判北京的胡同项目改造，但仅就外表来看，很多项目似乎一直陷入过度关注"模样"而忽视了本质的误区。

要知道，城市不是明信片，城市是生活的载体。

中国的城市和建筑经历了高速的发展，当下急需全社会和各行各业去试图脱离这个本质的误区，大家更喜欢新颖的、亮眼的设计来满足空间本质之外的某些需求，但当大家陷入某种审美疲劳后，新的形式和手法就不得不被逼迫出来以适应经营者的求新诉求。

这本书里，介绍了很多位于老旧城区内的项目，在优客工场和共享际团队的策划改造下，无论是废弃的酱油厂，还是胡同里的老宅院，都可以挖掘出令人惊讶的潜力。

在我看来，优客工场与共享际所参与的城市更新项目，是城市开发从"投资"到"使用"空间的转变，一个我们虽然还未介入但迟早都要深入了解的新阶段。

共享际和优客工场提早迈出了一大步。

作为一名建筑师，我对于空间的理解是这样的：空间本身的年代感并不取决于绝对意义上的时间序列，而更多地来自改造过程中的理念迭代和内容重塑，这从来不是一件容易的事。

我们在实践中会关注人在空间中的行为，但却对不同形态的行为缺乏深入的认识和理解，我们会关注空间带给人的直接情绪，却极少考虑过内容在空间中塑造的情绪价值，因情绪价值和认同而发生的传播、分享、聚集等间接行为，这需要在空间的设计中顾及品牌与内容的调性与性格，顾及因为这样的性格而聚集的人群的喜好，这样一个全新的思考链就让设计的实现有了出处和落脚点，而贯穿始终的正是"人"——经营的人和来访的人，他们都是这一个空间的参与者，也是内容的创造者。

我特别惊讶于这本图册中所收录的改造项目之多、品质之高，更令人难以置信的是改造速度之快。

优客工场和共享际的团队如何做到这一点?这本身就是一个非常值得研究的课题。我相信，在这项事业的背后，一定有一种年轻和震荡的能量，在为这个团队输出着巨大的动力。

美国波士顿华裔女市长吴弭的竞选口号是：每一个人的城市。这是一个多么浪漫的理想，也何尝不是每个人的期望。

共享际和优客工场所从事的城市更新事业，就是通往这一目标的努力与尝试。只有关注城市中"每一个人"的需求，才能拥有更多包容与创新的内容载体，基于此，以建筑空间的内容及其外延复兴城市文化，才有可能不断书写更加新颖的创意。

李 虎
OPEN建筑事务所创始合伙人

推荐序二

超级目的地
更加流动的场景如何以人为本

　　摆脱移动性和流动性的缺失，依托文化历史的在地性改造，因势利导的氛围营造，现代与未来感的交融碰撞，越来越多空间商业凭借自成一体的内容企划能力实现与用户的低成本和高信任连接，空间因此成为场景的容器、体验的代名。

数字化进程不仅繁荣"高像素"复刻的线上，也重构着个体对实体商业的理解与想象。在到家解决方案与同城物流效率成为基础设施的当下，人们依然在追求生活中距离更近、感受更近的亲密场景，本质是"体验式连接"与"数字化温度"。无论KOC社群中心或者同城精品酒店，空间成为"自我表达"的介质，场景绝非静态的物理存在，而表现为更加流动的内容与氛围。

王小波曾说，"有趣是一个开放的空间，一直伸往未知的领域"。今天用户复杂多元的生活方式追求尤其在疫后重新生长，这对空间商业的想象与实践提出了更高的要求。摆脱移动性和流动性的缺失，依托文化历史的在地性改造，因势利导的氛围营造，现代与未来感的交融碰撞，越来越多空间商业凭借自成一体的内容企划能力实现与用户的低成本和高信任连接，空间因此成为场景的容器、体验的代名。

大庆兄作为共享际和优客工场创始人，一直以"空间+内容"的运营思路持续实践新物种的创新，为空间商业的体验融合提供了理想可行的参照范本。在《城市更新空间内容探索》一书中，我们看见不仅是独出心裁的空间设计、内容企划，更多是一个个蕴含社群、交互、观念、温度的"智慧场景"，让"超级目的地"成为可能，让线下、附近有了新的打开方式。

社交网络无限放大社群的影响力与覆盖范围，也意味着用户信任养成对于社群运营的可持续价值。以实体门店为基础的"KOC社群中心"打造，承载了社群维系和交互的品牌客厅、城市新公共空间功能，让空间商业找到自己作为社交与社群关系的"容器"价值，以人格IP连接社交温度的大庆朗读便是例证。

当办公、娱乐、居住边界消解，当阅读、咖啡、策展成为新空间标配，从复合业态到融合坪效，是空间的充分利用，更是用户体验的场景融合。每增加一种场景解决方案，对消费者而言就多一分吸引力。无论是将居住、文化、工作、艺术叠加交互的东四·共享际，还是融合饮食、聚会、阅读、旅居的念念行旅，都凭借场景的融合与流动，成为新生活观念的策源地。

一切"未完成"的创设，都期待用户参与激活新的内容共创。与泡泡玛特的盲盒机制一样，空间商业也需要立足留白能力，让更多用户能够参与品牌创造和IP塑造。书中诠释的立足露营创建"有机生活"的无瓦农场，便为当下亟需的"附近的远方"提出了颇具体验感与仪式感的生活提案。

在线上繁荣的时代探索线下繁荣，势必要求空间商业以更加独特、有温度的体验成为"超级目的地"，然而作为目的地的商业只是过程与手段，"打卡运动"并非最终诉求，通过内容与氛围方案满足用户的心灵慰藉、意义找寻，常来常往，万象更新，才是时代命题所向。亲密场景时代，城市更新与空间内容探索大有可为，期待看见更多更流动的场景，成为真实生活的"超级目的地"。

吴　声

场景实验室创始人

前　言

经过改革开放40余年，伴随经济腾飞，中国的城市化进程远超发达国家。经济与城市化的高速发展，使得大量既有建筑难以满足当今人们的物质、文化和消费需求，成为低效资产，不能匹配区域经济发展所带来的资产价值，在国家调控政策的背景下，提升低效资产活力，更新城市空间，活化老旧城区，正逐步上升为国家战略。

共享际与优客工场在成立之初，就将既有建筑和低效存量资产作为了最主要的产品研发土壤。我们专注于以内容为核心的城市"空间+内容"运营，通过酒店、餐饮、运动、办公、居住、阅读、戏剧等多元内容的整合，创造出了一个新时代背景下面向新青年消费群体的多元生活方式集合。我们始终将新消费孵化、新场景融合与物理空间的更新紧密结合，通过新消费业态的挖掘、培育与孵化，定制满足新青年消费群体多元需求的消费场景，从经营内容出发，由内而外地进行城市与建筑空间的更新，从消费内容定制出发，由人的消费向空间的场景体验迁移，开展更新改造提升，开创可持续经营，切实改变消费生活品质的城市更新模式。

我们以运营为核心，从社群出发，在开展具体项目时，首先需要根据项目特点和环境，发掘客群的多元性和市场痛点，据此选定商业内容，构建产品，为特定区位、特定条件的建筑空间，注入定制化产品，实现对物理空间的提升与改造。

我们对运营的关注和探索也吸引了众多优秀合作品牌和空间，对设计的精益要求也结识了许多优秀和专业的设计团队。在这里，我们希望能将更多更好的内容和设计带给更多热爱生活的人，让城市充满活力，生活充满乐趣。

Preface

After more than 40 years of reform and opening up, China has witnessed rapid economic development. Along with the economic take-off, Chinese urbanization process is far faster than that of developed countries. As economy and the rapid development of urbanization, a large number of existing buildings is difficult to meet the physical, culture and consumer demand of today's people. A large number of existing buildings becoming inefficient assets, can't match the value of the asset brought about by region economy development. Under the background of the national regulation policy, enhance vitality of the inefficient assets update urban space, and activate the old town, is gradually rising to a national strategy.

At the beginning of its establishment, 5Lmeet and Ucommune have taken existing buildings and inefficient assets as the most important soil for product research and development. We focus on the content based urban "space + content" operation, through the integration of hotel, catering, sports, office, residence, reading, drama and other diverse content, to create a new era background for the new youth consumer groups of diverse lifestyle collection. We always will combine the new consumer hatch, new scenarios closely with the physical space of urban renewal.Through new forms of consumption of digging, cultivation and incubation, customized consumption scenarios of diversified demand to meet the new youth consumer groups, starting from the management content, update the architectural and urban space, starting from the content of consumption custom custom, From human consumption to space scene experience, carry out renewal, transformation and upgrading, create sustainable management, and effectively change the quality of consumer life urban renewal mode.

We take operation as the core and start from the community. When carrying out a specific project, we first need to explore the diversity of the customer base and the market pain points according to the characteristics and environment of the project, select commercial content, build products, and inject customized products into the building space of specific location and specific conditions to achieve the improvement and transformation of the physical space.

Our focus and exploration on operation has also attracted many excellent brands and spaces for cooperation, and we have met many excellent and professional design teams for lean design requirements. Here, we hope to bring more and better content and design to more people who love life, so that the city will be full of vitality and life will be full of fun.

模式创新

新消费社群挖掘

当今社会，Z 世代成为最具活力和消费潜力的年轻群体，他们更热爱分享，注重消费的社交属性，更愿意为爱好投资，消费群体呈现明显的圈层化，崇尚品质也注重颜值，网红与裂变传播成为重要的营销手段。社群的线下聚集需要空间具备新消费群体的主要需求，我们孵化和联合的流量商业品牌聚集大量线上活跃用户，赋能线下空间，为新消费群体打造全新消费场景和共享生活社区，通过线下活动和优质内容，提升社群黏性，推动裂变式社群营销，从而利用内容组合与分发，打造优质线下空间和垂直社群。

创新产品体系建构

与传统的开发思维下产品的标准化有非常大的区别，我们的产品核心是分散的、独立的内容和品牌，随着共享际的持续经营，以及与各类业态的不断合作发展和品牌孵化，开始构建起以休闲农业、职住一体、文旅酒店、戏剧工坊、城市民宿、社区商业为主的产品架构，在这些基本产品架构下，是共享际品牌下聚集的众多符合新青年消费群体需求的内容矩阵，丰富的内容矩阵和商业品牌支撑着产品从策划定位到实现落地过程，但这些产品并不是完全固定的组合，产品架构只限定产品的基本逻辑与经营方式。

订阅场景孵化品牌

线下空间区别于线上消费，最大的优势就是能够为来到线下的客人提供丰富的场景体验和互动方式，我们与每一个初创和成熟品牌的合作，不是单纯的租赁关系，而是与品牌共同从客户研究，产品定位，消费洞察，直到最终的空间实现，不断尝试和打磨品牌的新内容、新玩法，充分利用空间特质，创造全新的消费场景，让空间与品牌产生更多可能，提升用户体验，创造更多元的流量入口。在经营中不断探究流量与社群的场景数据，更新场景内容，不断提升品牌活力，为社群推送创造性新内容，激活社群活力，在品牌与社群间建立良好的互动关系。

Excavating new consumer community

In today's society, generation Z has become the most dynamic and most consumption potential of the young group, they are more love to share, pay more attention to the social attributes of consumption, more willing to invest for hobbies, consumer groups show obvious circle, advocating quality but also pay attention to the appearance, Internet celebrity and fission dissemination has become an important marketing method. Offline community gathered need space with the main demand of new consumer groups, the flow business brand we hatch and combine with gathered a large number of active users online, energized the offline space, build a new consumption scenarios and Shared life community for the new consumer groups.Through offline activities and premium content, we improve community viscidity, promote the fission community marketing, thus through using the combination and distribution of content, create high-quality offline space and vertical community.

Construct innovative product system

Which has very big difference between product standardization under traditional thinking, our product core is dispersed, independence and the content of the brand, as the 5Lmeet continuing operation, as well as with various formats of cooperation development and brand incubation.We have begun to build product structure of leisure agriculture, co-working and living, traveling hotel, theater workshop, city home stay, the community commercial. Under these basic product structure, it is the content of the matrix gathered under the 5Lmeet brand that can meet the demand of new youth consumer groups,the rich content of matrix and the business support the product from the planning of brand positioning to achieve landing process, but the combination of these products are not completely fixed.Product structuret only restrict he basic logic the mode of operation of product.

Subscribe the scene, Incubate the brand

Offline space different from online consumption, the biggest advantage is able to supply to offline guests with rich experience and interactive way.Each cooperation we take with start-up and mature brand, is not a pure rental, but new content and new gameplay from the customer together with the brand research, product positioning, consumer insight, until the final space, constantly try and polish the brand, make full use of the characteristics of space, create a new consumption scene, make space and brand more possiblities, improve user experience, create more diversified flow entrance. In the operation, we constantly explore the scene data of flow and community, update the scene content, and constantly improve the brand vitality, create new content for the community, activate the community vitality, and establish a good interactive relationship between the brand and the community.

空间 + 内容

空间

老建筑改造、城市更新项目与新建项目最大的区别在于空间不是结果，而是条件，不再是"从 0 到 1"，而是"从 1 到 100"的过程。因此在以往所有的设计与改造中，对于空间的阅读和理解是打造一个产品的最大挑战，文化和历史元素的保留，破旧结构的加固，墙体的拆除，空间结构的重塑。设计师往往需要在实现过程中"顺势而为""因地制宜"，虽然难度极大，但在阅读老建筑的过程中，无形地与一座几十年甚至上百年的建筑进行了一次穿越时空的对话。

老建筑的空间不仅仅局限在方寸之间，围墙之内，我们更多的时候会在意建筑所处的城市街区、地段，以及周边的生活。城市更新不同于新建，老建筑所处的位置，周边已经拥有了完善的生活设施和文化圈层，身居附近的居民已经形成了特有的生活方式，成熟的商业和文化发展会给一个区域标记出其特有的生活和文化印记。设计需要结合并利用这些生活和文化的印记，让设计更具有文化内涵和扎实的生活感知。在这些空间中，设计最本质的追求是人在空间中的行为，既得体、舒适，又惊喜连连，碎片化的空间体验带来完整的愉悦体验，串联这些的，往往是老建筑自身的深厚场景感，这也正是城市更新、老建筑改造的魅力所在。

内容

老建筑和存量资产的闲置，往往伴随着周边产业发展不足，人口消费活力低下等问题，不论是在老城区还是新建城区，没有产业和消费的拉动难以实现项目更新改造后的持续经营，因此我们在项目的策划、设计、招商、运营阶段更加注重具有内容属性的商业和业态类型。办公和公寓引入保障性人流，支撑商业起步，同时在商业配比中兼顾流量品牌和大众消费品牌，利用区域人流，自有人流，目标性人流为商业活力的提升与恢复提供可靠保障，提升项目活力。

好的内容，能够持续经营，不断更新，自带流量，良性发展，打通线上线下，居住、办公、餐饮、戏剧、阅读、运动等垂直内容的注入，提供一站式城市生活驿站，关注空间中人的消费与经营，在可持续的运营中孵化内容，为更加广泛的城市更新提供新模式与新内容。

Space and Content

Space

The biggest difference between urban renewal projects and new projects is that space is not a result, but a condition, no longer "from 0 to 1", but "from 1 to 2" process. So in the past all of the design and transformation, the space of reading and understanding are the biggest challenges, to create a product of cultural and historical elements, dilapidated structure reinforcement, the removal of the wall, the reconstruction of the spatial structure, designers often need to be in the process of implementation "go with the flow", "adjust measures to local conditions", although difficult, but in the process of reading of the old buildings, it seems like making a time-travelling dialogue invisibly with a building that is decades or even hundreds of years old.

The space of old buildings is not only limited to the square inch, within the wall, we will pay more attention to the city block, where the building is located, and the surrounding life. Urban renewal is different from new construction. The old buildings have perfect living facilities and cultural circle around them. The residents living nearby have formed a unique life style. Mature commerical and cultural development will mark a region with its unique life and culture. Design needs to combine and use these life and cultural imprints, so that the design has more cultural connotation and a solid sense of life. In these spaces, the most essential pursuit of design is the behavior of people in the space, which is decent and comfortable, and surprises repeatedly. Fragmented space experience brings complete pleasure experience, which is often connected by the profound scene of old buildings themselves, which is also the charm of urban renewal and old building renovation.

Content

Old buildings and the existing assets, often accompanied by peripheral industry development, the population problem, low energy consumption and so on, both in the old and the new urban area, without industry and consumption of drive is difficult to realize project after upgrading to continue as a going concern, so we in the project planning, design, development, operation stage pay more attention to business and formats with content attribute type. At the same time, both flow brands and mass consumer brands should be taken into account in the commercial ratio. Regional flow, self-owned flow and target flow should be used to provide reliable guarantee for the promotion and restoration of commercial vitality, so as to enhance the project vitality.

Good content, can be continuous operation, constantly updated, bringing traffic, benign development, through online and offline, residential, office, restaurant, theater, reading, sports, such as vertical injection content, provide one-stop station city life, focus on the human consumption and space management, and content in the sustainable operation of incubation, providing more extensive urban update new mode and new content.

空间内容的设计

内容的构想来自空间的启迪，空间的更新来自内容的活力。两者密不可分，相辅相成，在项目的更新与运营周期中两者都彼此成就，因此我们在任何项目中，都会将内容的定制与研究放置在最首要的位置上。根据不同的项目类型选择适合的设计师进行合作，通过与大量的品牌、内容和设计师合作，我们建立了设计师 + 内容的合作方式，未来还会有更多更好的内容与空间经过设计师的创意呈现，也会有更加成熟的空间与内容组合方式，完成更好的城市更新项目。

Design for Space and Content

The idea of the content comes from the enlightenment of the space, updated vitality comes from the content of the space. Both inseparable, supply each other, both in the project updates and operation cycle achieve each other, so in any project, content customization and research will be placed in the first place, according to the different project types to choose suitable for designers to cooperate, Through the cooperation with alarge number of brands, content and designers, we have established the cooperation mode of designer + content. In the future, there will be more good content and space through the creative presentation of designers, and there will be more mature combination of space and content to complete better urban renewal projects.

Contents

居住空间

职住空间

办公空间

新消费空间

居住空间
Residential Space

酒店、民宿、民居建筑

关于空间

居住建筑空间包含了人在空间中最多样丰富的行为状态，在相对较小的室内空间中，行为尺度和空间转换都变得更加敏感，设计需要适度地提炼空间的精髓，专注行为的便捷，在融入更多新场景、新元素后，让人获得的是更加舒适、更多趣味及更加利于传播的空间特质。

关于运营

居住空间的运营，需要结合空间设计，赋予空间相应的精神内涵，如民宿彰显本地性，亲近如朋友般的关系，在传统街区的新空间体会本地生活最真实的场景。酒店则需要根据客群建立良好的社群互动，营造标志性的空间，吸引流量，打造差异化品牌。

Hotel and Residential House
Space
Residential building space contains a person's most abundant behaviors, in the interior space which is relatively small, there will be more sensitive behavior scale and spatial transformation, the essence of design requires moderate refining space, focus behavior's is convenient, to blend in more new scene and new elements, let a person get more comfortable, more fun, and more conducive to communication.

Operation
The operation of the living space needs to combine with the space design and give the space corresponding spiritual connotation. For example, the homestay reflects the local nature, and the close relationship is like a friend. The most real scene of local life can be realized in the new space of the traditional block. Hotels need to establish good community interaction according to customer groups, create iconic space, attract flow and create differentiated brands.

念念行旅

项目背景

念念行旅位于北京市东城区东四北大街 69 号，整栋建筑是 1975 年建成的北京市晒图厂（印制制造和建筑业施工依据的蓝图）厂房，后经几次改造，改为酒店经营多年。建筑外观地上 4 层，总面积 3000 平方米，为多层混合结构（外墙与内部柱梁承重）。主要由杂志书店与酒店大堂，四个商户，三层客房和一个 500 平方米的观景露台组成，除客房外，公共空间之间相互穿插，室内室外多样的动线组织，形成一个相互连通，内容丰富的奇妙领域。

中国·北京 东城区东四北大街69号

Project Background

Read and Rest Hotel is located at No. 69 Dongsi North Street, Dongcheng District, Beijing. The whole building is the factory building of Beijing Printing Factory (a blueprint for printing manufacturing and construction industry construction basis) built in 1975. After several renovations, it has been transformed into a hotel for many years. The building has four floors above the ground, with a total area of 3000 square meters, and is a multi-storey mixed structure (external walls and internal columns and beams). The main functions are the magazine and bookstore and the hotel lobby, four merchants, three-floors guest rooms and a 500-square-meter viewing terrace. Except for the guest rooms, the public spaces are interspersed with each other to form a variety of indoor and outdoor moving lines, forming a mutual Connected, content-rich and wonderful areas.

城市语境

北京的胡同里偶尔会出现几个高耸的楼房,都有自己的历史和来由。城市肌理和胡同生活方式也有着不同的改变,念念行旅所处的位置商业氛围更加浓厚,生活气息则不是很强,我们希望通过这个项目能够为胡同带回一些原本居民的生活氛围和老城区的街区秩序。

Urban Context

Occasionally, several towering buildings appear in Beijing's Hutong, each with its own history and origin. The urban texture and the Hutong lifestyle have also changed differently. The location where Read and Rest Hotel is located has a stronger commercial atmosphere, but the living atmosphere is not very strong. We hope that through this project, we can bring back some of the original residents' living atmosphere and the order of the neighborhood in the old city.

设计理念

设计将胡同、院落、屋顶这些空间元素，通过视觉、材质触感等方式，引入一层的空间。

我们将胡同空间自然地引入室内，创造了一个遮风避雨的室外露台。在露台上，通过建筑原有的洞口，犹如中式园林中的框景，呈现了一幅层叠起伏的北京民居画卷。

建筑北侧外墙与院墙间的狭窄空隙被改造成一个灰砖墙面的小庭院，院内布置定制的盆景绿植，有两个宽大的落地窗，可以在室内观赏庭院中的别致景色。

建筑的首层植入小咖啡厅空间，与室内空间自然地融合，同时向胡同内打开的窗口，创造一个新的街道行为空间和消费场景。

Design Concept

The design introduces spatial elements such as Hutongs, courtyards and roofs into the space of the first floor through visual and tactile methods.

We bring the Hutong space naturally into the interior, creating an outdoor terrace to shelter from wind and rain. On the terrace, you can pass through the original hole of the building, just like the frame view in Chinese garden, presenting a cascading picture of Beijing folk houses.

The narrow air raid between the north facade of the building and the courtyard wall has been transformed into a small courtyard with gray brick walls. The courtyard is decorated with custom bonsai plants and has two large floor-to-ceiling windows to enjoy the unique views of the courtyard.

The first floor of the building is implanted with a small cafe space, which naturally integrates with the interior space and opens a window into the alley to create a new street behavior space and consumption scene.

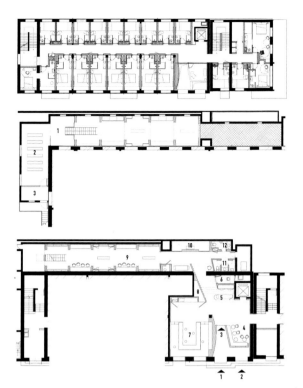

1 大堂
2 活动空间
3 储藏间

1、2 入口
3 主入口
4、5 休息区
6 前台
7 办公区
8 等候区
9 储藏间
10 院子
11 休息区和储藏间
12 卫生间

• 我们希望在建筑的首层空间创造不同形式的阅读空间，进入正门，左侧宽敞、明亮的空间是一个用于接待客人和举办活动的大客厅。围合而坐的沙发，外侧微微抬起的地台，增加了空间的层次。在容纳更多客人的同时，也像一个展厅，展示着墙面的杂志、玻璃窗外的胡同生活、陈设的展品。南侧高大纯净的玻璃窗，是向城市展示阅读生活的一个橱窗，是了解这个杂志般空间的封面。

• 我们在前台边设计了一个大大的方形窗，坐在窗边，看着胡同里的人来人往，成为阅读城市的一个窗口。

• 沿着波浪般起伏的灰色墙面向深处走去，橱窗中别致静雅的小院散发着中国古典园林的禅意和静谧，引人入胜。

• 我们将书店原址的二层建筑楼板拆除，改造成了一个两层通高的空间，进入书店后，空间的瞬间开阔给人更加松弛和自由的感受，富有韵律的室内空间分割出了若干使用单元，分布书架、座位和展台空间。

• We hope that in the construction of the first floor space to create different form of reading space, enter the main entrance, on the left side of the capacious and bright space for guest and host activities is a huge living room, surrounding the sofa, sitting with the platform of the lateral slightly raised, increased the level of the space, to accommodate more guests at the same time, also like an exhibition hall, showing the magazine of metope, Hutong life outside the glass window, display exhibits. The tall, pure glass window on the south side is a window to the city showing the life of reading, a cover to understand this magazine-like space.

• We designed a large square window in front of the reception desk. Sitting by the window, you can watch the people coming and going in the alley, which becomes a window to read the city.

• Walking along the undulating gray walls, the unique and elegant courtyard in the shop window exudes the zen and tranquility of classical Chinese gardens, which is fascinating.

• We removed the two-story floor of the original site of the bookstore and transformed it into a two-story double-height space. After entering the bookstore, the space opens up instantly, giving people a more relaxed and free feeling. The rhythmic interior space is divided into several use units, including bookshelves, seats and booth space.

Interior and Soft outfit Design

Read and Rest Hotel is located in Beijing's downtown and old city. At the same time, reading is the theme element. In the soft decoration design, the modern wabi style of "WABI SABI" is used to create a quiet indoor environment in the noisy, while echoing the concept of spiritual wealth in the reading culture. The soft outfit of Read and Rest Hotel adopts more concise and elegant colors and materials. Gray-white Swedish imported fabrics, logs, rattan, and Italian imported cowhide constitute low-key and textured furniture elements. In the selection of products, the main focus is the concept of "one person study", which is based on the selection of single products, such as the TON dining chair in the lobby lounge area, the chief chair in the double room and the exquisite Stellar Works Chillax single sofa in the suite, enhances the sense of design while creating a comfortable personal reading environment.

室内及软装

念念行旅酒店地处北京闹市老城区，同时又以阅读为主题元素，在软装搭配上设计选用了"WABI SABI"的现代侘寂风格，打造闹中取静的室内环境，同时呼应阅读文化中的精神富足理念。念念行旅的软装搭配上采用了更简洁、典雅的色系和材质，灰白瑞典进口面料、原木、藤编、意大利进口牛皮等材质组成低调且具有质感的家具元素。在选品上主打少而精的"一人书房"理念，通过单品的选用衬托念念行旅的整体气质，如大堂休息区的 TON 餐椅、大床房的酋长椅和套房内精致的 Stellar Works Chillax 单人沙发，提升设计感的同时打造舒适的个人阅读环境。

挑　战

- 小尺度酒店空间的功能完备与功能复合
- 如何打造丰富空间形式和内容需求的阅读场景
- 书店空间的取舍与重塑
- 如何更好地在胡同社区中打造既融合又具有辨识度和吸引力的空间
- 如何在旧厂房内植入新功能，满足新业态的空间需求，同时保证结构安全和设计完成度

Challenge

• Complete function and composite function of small-scale hotel space

• How to create a reading scene with rich spatial forms and content needs

• Selection and reshaping of bookstore space

• How to better create a fusion, recognizable and attractive space in the Hutong community

• How to implant new functions in the old factory building to meet the space requirements of the new format, while ensuring structural safety and design completion

视觉系统

运营者说

王翔_共享际设计总监/首席建筑师

念念行旅并非传统意义上的酒店，设计之初我们就希望赋予这个空间更多的属性和场景内涵。胡同本身所具备的在地生活精神内核给予了这栋建筑更广阔的延伸性，酒店的公共空间包容着更广泛的公共性，让阅读、社交、品牌呈现、新生活的展示成为一个个触手，抓取更多元的场景，也成为胡同居民体验新文化和新生活的窗口。

吴雪_共享际首席室内设计师

念念行旅酒店是对旧建筑的重新定义，通过设计重新切割及新的消费场景植入之后，让原本单一的传统建筑体，变成了互动性高、新场景、内容丰富的设计时尚酒店，是设计加运营共同产生的化学反应。

项目信息

项目名称：念念行旅酒店

地点：北京市东城区东四北大街 69 号

内容策划：优享与行策划咨询有限公司

甲方团队：王翔，钟宇辰，吴雪

设计公司：Office AIO

设计师：Tim Guan，Isabella Sun

设计时间：2018 年 11 月～ 2019 年 3 月

竣工时间：2019 年 9 月

项目面积：3000 平方米

隐 院

项目背景

北京的胡同，不仅仅是一种特有的居住方式，更是很多人儿时曾经的记忆。随着现代化与城市化的进程，许多人搬离了胡同，但是胡同生活的那一份烟火气，却永远地留存在了记忆里。

岁月给人也给建筑留下了时光的印记，老去的房子、腐烂的柱子、杂草丛生的院子……胡同的生活模式，在城市发展进程中不断地"被动地"消失着。这种被动导致的是人们记忆的断层，仿佛儿时的回忆还在昨天，但是今天却已经生活在现代化的建筑中，人与人的隔阂、人与建筑的隔阂也越来越大。如何把这种被动产生的消隐重现、让历史与美好的记忆重新回归，并且让更新后的建筑与自然环境结合，产生一种历史与当下的对话，是我们在这座胡同小宅改造中一直探讨的问题。

改造并不是对老建筑的"维修"与"复刻"，而是要将一种新的生活方式与旧有的历史叠加，并且产生新的对比与融合。

中国·北京 东城区宝钞胡同90号

Project Background

The Hutong in Beijing is not only a lifestyle but also the memory of many people's childhood. With the process of modernization and urbanization, many people have moved out of the Hutong, but the feelings of Hutong life remain in one's memory forever.

Time has left the mark on buildings, old houses, rotten pillars, and overgrown yards. Hutong's lifestyle has been "passively" disappeared in the process of urban development. This kind of passivity leads to the disorder of people's memory, as if childhood memories are still yesterday, but today they are already living in modern buildings, and the estrangement between people and buildings is increasing. How to reappear this passive concealment, return history and beautiful memories, and combine the updated architecture with the natural environment to produce a dialogue between history and the present? It is a problem that we have been trying to explore at the beginning of the renovation of this hutong house.

Renovation is not the "maintenance" and "reproduction" of the old buildings, but to superimpose a new lifestyle with the old history and create a new contrast and integration.

重现与再消隐

这座小院是业主儿时与爷爷奶奶一起生活居住的房子，由于多年在外漂泊，房子已经破败不堪，小时候的种种回忆也随之而去。如今想要搬回到胡同中生活，但是房子的现状已经不适合居住。此次改造的目的，就是要给业主营造一个适应现代生活方式的居住空间，并且唤醒曾经"被消失"的那些胡同生活印记。

Reappear and Disappear again

This small courtyard is the house where the owner lived with her grandparents when she was a child. Due to years of wandering, the house has been dilapidated, and all kinds of childhood memories have gone with it. Now she wants to move back to Hutong. But the current situation of the house is no longer suitable for living. The purpose of this renovation is to create a living space suitable for the modern lifestyle for the client, and to awaken the life marks of those Hutongs that have been disappeared.

镜与万花筒

在院子的地面与墙面上，采用了镜面玻璃，材料特有的反射性可以让周围的胡同、树木、天空都映射到院子的地面中，给空间带来了更多的可能性，也让这种虚实的互动性更加凸显。人在景中、人在镜中，在层层叠叠的反射与通透的对景之中，给人一种如梦如幻的空间体验感。

镜面反射的自然环境每时每刻都在变化着，与镜面组合形成了一种类似于"万花筒"的效果。万花筒的英语名称 KALEIDOSCOPE，集合了希腊语的 KALOS（美丽）、EIDOS（形状）和 SCOPE（观看）等词汇，其实也是概括了万花筒的几大特点。

随着反射与周围环境的不断变化，镜面的小院展现出了一种瞬息万变之美。每一个瞬间都是独一无二的，就像时间一样，不能被记录，只能被感受。院子中这种"有形"反射出的"无形"的状态，也让观者与环境、环境与建筑、建筑与人之间都产生一种对视的关系。这种关系消隐了空间的边界，也消隐了人与空间之间的距离感。

Mirror and Kaleidoscope

Mirror glass is used on the ground and walls of the yard. The unique reflectivity of the material allows the surrounding Hutongs, trees, and sky to be reflected on the ground of the yard, which brings more possibilities to space. This kind of interaction between virtual and reality is more prominent. People in the scene, people also in the mirror, in the layered reflection and transparent contrast, give people a sense of space experience like a dream.

The natural environment of specular reflection is changing all the time, and the combination with the mirror forms a kind of effect similar to "kaleidoscope". The English name KALEIDOSCOPE is a collection of Greek words such as KALOS (beautiful), EIDOS (shape), and SCOPE (watch). In fact, it also summarizes several major features of the kaleidoscope.

With the continuous changes of reflection and the surrounding environment, the mirrored courtyard shows a rapidly changing beauty, each moment is unique, just like time, cannot be recorded but can only be felt. The "invisible" state reflected by the "tangible" in the courtyard also makes the viewer and the environment, the environment and architecture, architecture, and people all have a kind of eye-to-eye relationship, which hides the boundary of the space. it also hides the sense of distance between people and space.

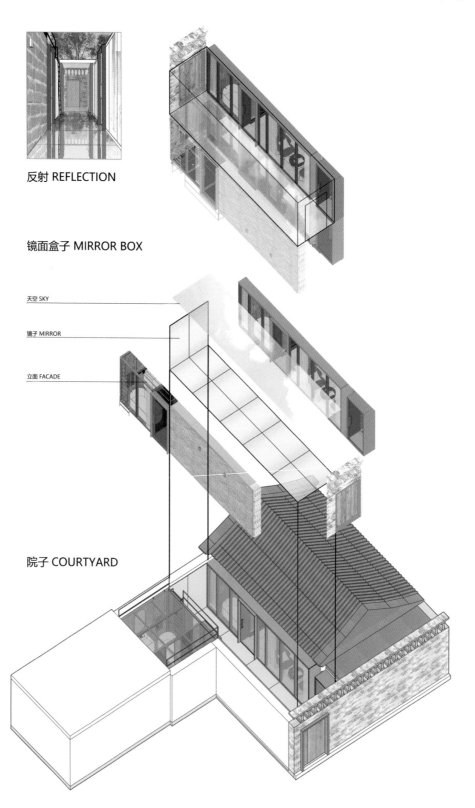

反射 REFLECTION

镜面盒子 MIRROR BOX

天空 SKY

镜子 MIRROR

立面 FACADE

院子 COURTYARD

反　射

院子仅有 12 平方米左右，呈狭长状，地面与墙面的玻璃镜面让小院的面积在视觉上得到扩大。当业主踏入大门的一瞬间，就会被镜面反射的透视效果包围，增加了空间的通透感，给人以双倍的景观体验。

Reflection

The small courtyard is only about 12 square meters, showing a long and narrow shape. The glass mirrors of the ground and walls enlarge the area of the courtyard visually. When the owner steps into the door, she will be surrounded by the perspective effect of specular reflection, which increases the permeability of the space and gives people double landscape experience.

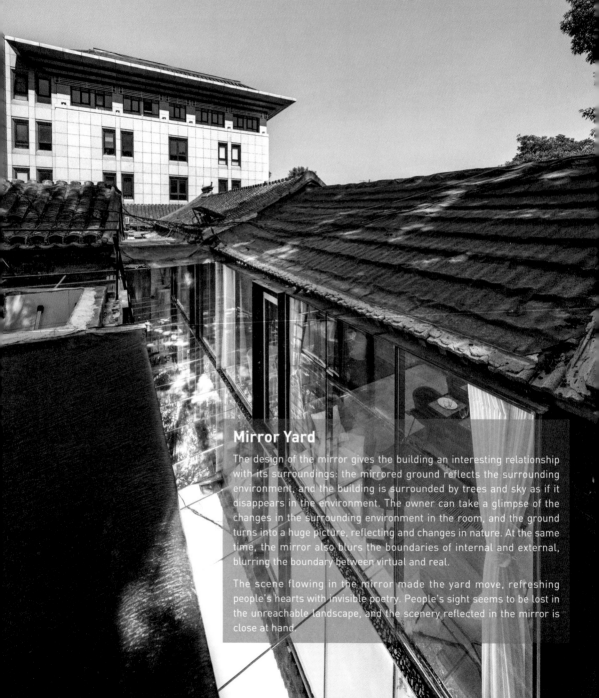

隐 院

镜面的设计赋予建筑与周围环境一个有趣的关系：镜面地面映射出周围环境，建筑被树木、天空所包围，仿佛消失在环境之中。业主可以在房间内一窥周围环境的变化，地面化作一幅巨大的画卷，映射出自然界的风吹草动、风云变幻。同时，镜面也模糊了内外空间的界限、模糊了虚实之间的界限。

在镜面上流动着的景象让原本静止的院子动了起来，以无形胜有形的诗意沁润着人们的心灵。人们的视线仿佛迷失在遥不可及的景观中，而反射在镜面中的景色又近在咫尺。

Mirror Yard

The design of the mirror gives the building an interesting relationship with its surroundings: the mirrored ground reflects the surrounding environment, and the building is surrounded by trees and sky as if it disappears in the environment. The owner can take a glimpse of the changes in the surrounding environment in the room, and the ground turns into a huge picture, reflecting and changes in nature. At the same time, the mirror also blurs the boundaries of internal and external, blurring the boundary between virtual and real.

The scene flowing in the mirror made the yard move, refreshing people's hearts with invisible poetry. People's sight seems to be lost in the unreachable landscape, and the scenery reflected in the mirror is close at hand.

通透空间

胡同住宅的改造过程其实就是现代生活方式与传统空间的一种融合。在改造中，将年久失修的木结构进行结构加固，将外立面改为玻璃幕墙的形式。通透的幕墙给室内空间增加了采光，让视线在院子两侧可以互相穿透，营造出对外内向、对内外向型的庭院空间。

Transparent Space

The transformation process of Hutong residence is a kind of integration of modern lifestyle and traditional space. In the transformation, the wooden structure which has been in disrepair has been strengthened, and the facade has been changed into the form of glass curtain wall. The transparent curtain wall adds daylighting to the interior space, allowing the line of sight to penetrate each other on both sides of the courtyard, creating an extroverted and introverted courtyard space.

1 入口
2 厨房
3 餐厅
4 洗手间
5 庭院
6 卧室

设计者说

申江海_建筑师/作家

大家一想到做景观，就是灰砖、灰瓦，然后再加点根雕、小桥流水什么的，但是我觉得这样挺没意思的。与业主一商量，觉得不如利用镜子，把小院做成一个万花筒。这个院子 12 平方米，特别小，做成镜子，那么每天反射的都是胡同周边的天空、树木，随着阳光的变化，镜子反射的场景也是变化的，整个院子就活起来了。

打磨厂 · 共享际

项目背景

项目位于北京前门的西打磨厂街，这是一条极具历史韵味的街道。打磨厂街形成于明代，清中叶起直至民国年间，繁盛一时，与西河沿、鲜鱼口、大栅栏并称为"前门外四大商业街"。中华人民共和国成立后大多数买卖商铺都已不复存在，古建筑变成了大杂院，但从残留的建筑物上仍依稀可见当年这条街的热闹与繁华。它们静默地等待着从历史中被唤醒，重新焕发出属于这个时代的光彩。

中国·北京 东城区长巷三条1号

Project Background

The project is located in Grinding Factory Street, Qianmen, Beijing, which is a street with a great historical atmosphere. Grinding Factory Street was formed in the Ming Dynasty and flourished from the middle of the Qing Dynasty to the Republic of China era. It was also known as the "four major commercial streets outside Qianmen " along with Xiheyan, Xianyukou, and Dashilan. After the founding of the People's Republic of China, most of the shops no longer exist, and the ancient buildings became large courtyards. However, from the remaining buildings, the prosperity of this street can still be seen in some parts. they are silently waiting to be awakened from history.

公共区域

公寓由三个院子相连而成，从一面拱形木质大门走进院子，左侧为前台接待区，右侧为中央厨房。厨房对着庭院里的古树，为了将庭院里的古树完好地保留下来，平面布局上巧妙地避开了古树的位置，用曲面的玻璃幕墙把厨房与庭院做了分隔。顺着接待区往里走，就是餐厅、酒吧。餐厅是一个弹性使用的空间，既可以作为公寓住户的用餐区，也可以作为一个举办公共聚会活动的地方。接待、中央厨房、餐厅和酒吧都采用了玫瑰金色的金属板，在这些公共空间里保留了建筑原本的砖墙，使酷炫、光滑、金黄的金属板与沧桑、粗糙、灰暗的旧砖墙形成气质上的反差，新与旧相互产生碰撞，拉开时间上的层叠。

室内地面采用光滑的灰色地砖，表面附有质感的纹理，庭院地面则由白色鹅卵石堆砌而成，内与外产生触感的变化。

1. 原始建筑布局

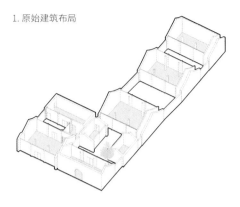

2. 加建玻璃幕墙

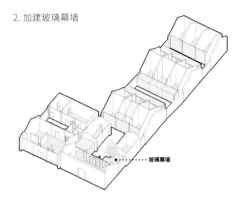

玻璃幕墙

3. 功能分区

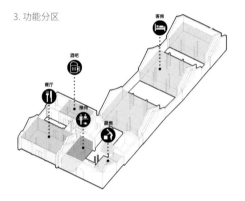

客房
酒吧
餐厅
接待
厨房

4. 改造后的建筑关系

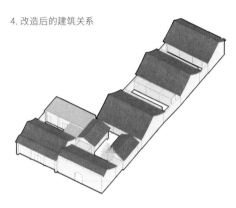

Public Area

The apartment is connected by three yards, with the entrance made of an arched wooden gate leading into the courtyard, which is divided into the front reception area on the left and the central kitchen on the right. The kitchen faces the ancient trees in the courtyard. These ancient trees are deliberately kept out of the plane layout in order to keep the them intact in the courtyard, and the kitchen is separated from the courtyard with the curved glass curtain wall. Walking along down the reception area, one could find a dining room and a bar. The dining room is a flexible space that can be used either as the dining area of the apartment's residents or a public place for gatherings. Rose-gold metal plates are used in reception, central kitchen, dining room and bar and the building's original brick walls are retained in these public spaces, making the cool, smooth, golden metal plates stand in stark contrast to the shabby, rough, gray old brick wall. This is indeed the collision between the new and old.

The smooth gray floor tile with finely textured surface is used in the interior floor, while the courtyard ground is piled by the white cobblestone, creating a different sensation of touch while walking from outside to inside.

中央厨房

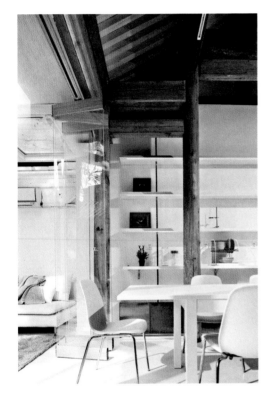

接待区

餐厅及酒吧

1 接待
2 中央厨房
3 洗衣房
4 设备间
5 锅炉房
6 餐厅
7 卫生间
8 酒吧
9 厨房
10 庭院
11 客厅
12 客房卫生间
13 卧室

客　房

穿过西侧的狭长走廊就是客房区域，利用原本的建筑结构，一个建筑体被划分为五到六间客房。玻璃自身的透明特性不会遮挡减损原有建筑物，不仅符合对阳光的追求，更能够从视觉上以及空间原理上使得原有的古建筑不会被削弱，所以整个院子的加建部分都是采用玻璃盒子的形式。再加上光线和人的运动，整个空间被赋予了活力。阳光穿过通透的玻璃倾泻到公寓内，给公寓内的一切带来温暖和灵动之感。

Guest Rooms

The long and narrow corridor on the west leads to the guest rooms. There are five or six guest rooms divided based on the original architectural structure of the building. The transparency of the glass means that it will neither block the much-desired sunshine nor impair the original look of the building, creating a visually and spatially balance in relation to the old building. Therefore, the extended part of the whole yard is built in the form of a glass box. With light flittering through and light and people moving around in it, the entire space is full of vitality. When sunlight pours through the transparent glass into the apartment, it brings warmth and energy to everything in the apartment.

运营者说

贾晓萌Nikky_共享际执行CEO/联合创始人

在共享际探索城市更新与内容创新的岁月和历程中，汲取了大量来自社会各界不同领域、不同行业、不同观点的朋友们和伙伴们的思考，作为创造每一个产品，实现每一个项目，创建每一个社群的根基。他们中有些与共享际走到一起，一同为新的内容努力与发声。

王翔_共享际设计总监/首席建筑师

共享际从关注人开始，筛选和孵化经营内容到卓越的空间打造，再到层出不穷的社群发酵，为消费人群创造价值认同的品牌与场景，真正做到了在商业消费市场中连接内容提供者与内容消费者，让线下空间不再难以经营，让更多元的消费群体乐于来到线下空间。

1 接待
2 洗衣房
3 庭院
4 客房

丫吉宿集

项目背景

"丫吉宿集"的故事就发生在一个以"桃花"而闻名的地方——北京市平谷区。每当春天到来，慕名前往的游客不计其数。除了桃花，平谷区还有著名的丫髻山风景区，因有道教庙宇坐落在山上，从古至今都受到皇室与百姓的敬仰。因此我们团队在最初设计"宿集"的时候，便决定"入乡随俗"，不仅要做到尊重当地的道教文化，也想要将当地的民风民俗作为一些特点融入设计当中。我们希望让"逃离"城市快节奏生活的人们，能够短暂地体验几天逍遥的隐世生活，当一次潇洒的"唐伯虎"。

中国·北京 平谷区前髻山村

Project Background

The story of "Yaji villageB&B Group" takes place in a place famous for "peach blossom" —Pinggu District of Beijing. When spring comes, countless tourists are attracted to it. Besides peach blossoms, the famous Ya Ji Mountain scenic area in Pinggu District, that Taoist temples are located on, has been respected by the royal family and ordinary people since ancient times. Therefore, when designing "B&B Group", our team decided to "do as the Romans do", not only respecting the local Taoist culture, but also integrating the local folk customs into the design as some characteristics. We hope that people who "escape" from the fast pace life in the city can experience a few days of carefree life, and become a natural and unrestrained "Tang Pak Hu".

设计理念

为了打造一个"完美"且"特别"的民宿，设计团队决定从材质入手。在一个完整的设计中，灵感与便捷等方面都是需要一些时间去感受的，只有"材质"是最直接与人们接触的项目，所以它的选择直接影响了客人们的居住感受。

外观方面选用了硅藻泥，这样可以"入乡随俗"，不会与周围的村宅显得格格不入，还可以加强设计团队所打造的"天然住宅"的理念。

室内的材料主要选用了肌理漆和木饰面的搭配。木饰面本就源于大自然，与肌理漆的搭配是相辅相成的。这个设计不仅"模糊"了房子的边界感，让人感到吃在自然中、住在自然中，还给人清新、舒爽的感觉，仿佛是一个让人"忘记烦恼"的神秘盒子。

最后，落地窗的大量使用，给房间带来了明亮的自然光线，到了傍晚也不用开很多电灯。想象一下，当我们站在窗边，举一杯散发着淡淡花香的桃花酒，满眼都是诗和远方，这不就是我们向往的生活吗！

Design Concept

To create a "perfect" and "special" B&B, the design team decided to start with materials. In a complete design, inspiration, convenience and other aspects need some time to feel, only "material" is the most directly contacting with people, so its choice has a direct impact on guests' living experience.

In terms of appearance, diatom mud is selected for "do as the Romans do" and that will not only incompatible with the surrounding village houses, but strengthen the concept of "natural house" created by the design team.

The interior materials are texture paint and wood veneer. Wood veneer originally comes from nature, and can be complementary to texture paint. This design not only "blurred" the boundary of the house, let people feel eating and living in nature, but also give people fresh and comfortable feeling, like a mysterious box to let people "forget the trouble".

Finally, the extensive use of floor-to-ceiling windows brings bright natural light into the room, which in the evening does not require many electric lights.Just imagine, when we stand at the window, holding a cup of peach blossom wine, our mind are full of poetry and distance, this is the life we yearn for.

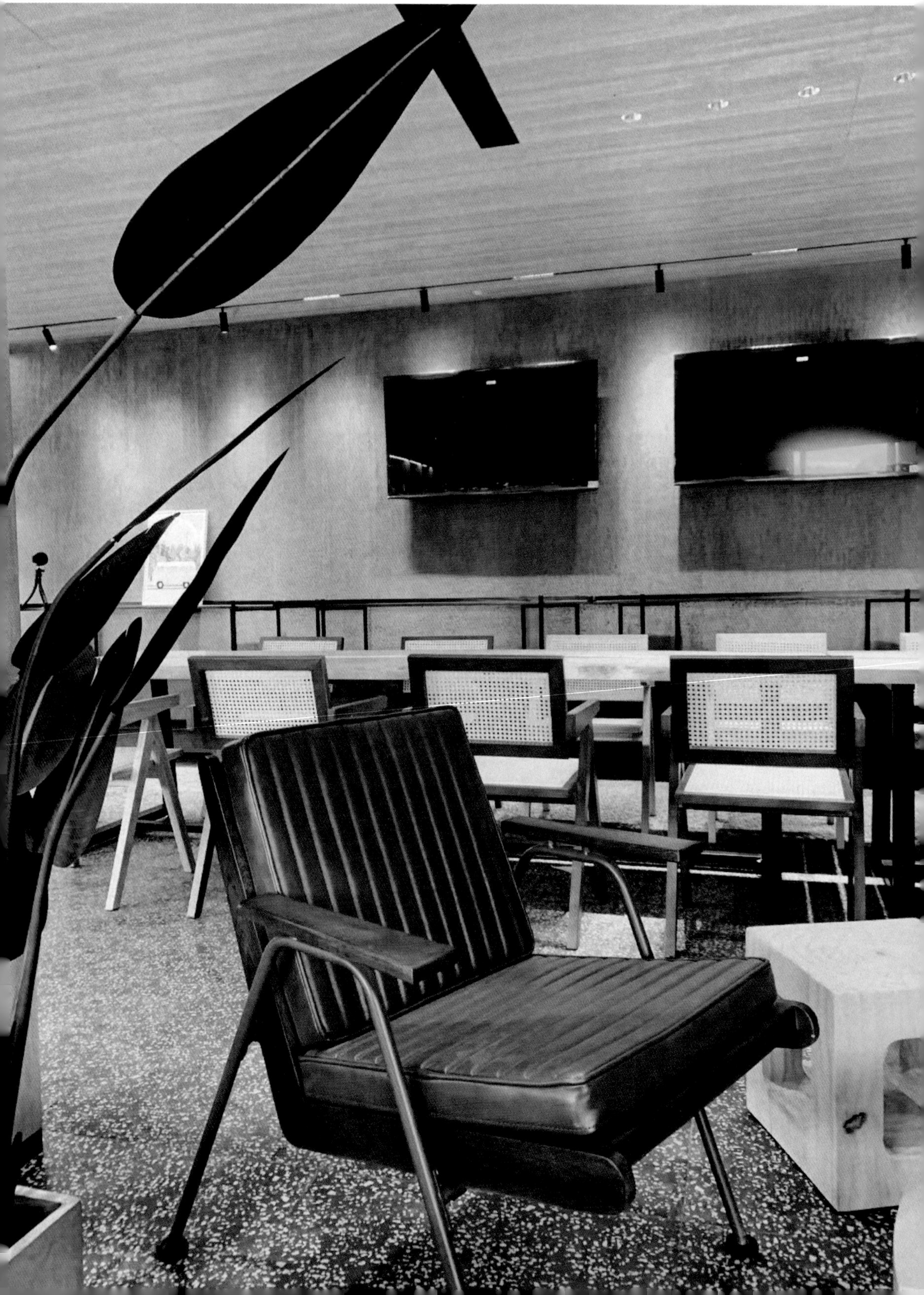

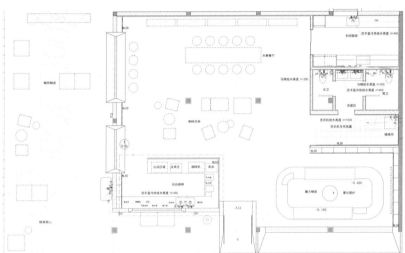

毛大庆

共享际/优客工场创始人、董事长

建筑师

丫吉宿集发起人

一座建筑，当它出现在一个生活气息非常浓郁的场景中时，它首先要呈现出的气氛必须是融合与友善的，而不应该是隔膜与冷峻。

乡村民宿的理想意义应当是缩小城乡差别，提升农村的美誉度和艺术鉴赏力，同时通过城里人在民宿的消费拉动农村消费，加强农民对民宿经营的参与度，力求把民宿变成一个"开门"的业务。

民宿，其实是新旧之间一种对话方式的建立，我们更多地是在强调宿集而不是民宿，是因为它应该是一个集群化的东西，是一个内容集合体，而不再是个案。

我们主张把微改造和乡土味这两个东西作为宿集发展的核心理念，那么人们就会形成一种真实的体验感、乡土感；同时在农民眼中，宿集又是能够给他们带来实际价值的，这才是一个大家互利共赢的局面，也可以拉动城乡之间人与人真正的流动跟融合。

所以，我认为这项工作的意义已经超越了建筑设计本身，也不是简单的空间规划，而是乡村文明再造的内容集合，真正的着力点其实应该是农村，是农民，是农业，如果忘记三农，片面地谈房子或者谈建筑设计，就会变成一个设计者运营的过程，这是有问题的。

乡村建设与更新面临着比城市中的改造更新更大的难题与挑战，城市中的更新可以依托庞大的在地人群，打造精准的产品形态，而远离人群聚集的乡村没有庞大的在地人群支撑，交通与配套设施又无法与城市相比，因此乡村的建设与更新就需要走出一条全新的经营之路，创新农业，专注体验，设计赋能，让逆城市化的人潮回到一个田园版的乡村。

愿　景

"宿集"这一概念开启了中国民宿集群时代，而"丫吉宿集"不仅是一个提供居住的场所，还融入了当地的文化，注重文创的发展。我们特别定制了展示架来展示当地的特色手工艺品，希望来访的游客在旅行结束时也可以带着专属的记忆踏上归程。

平谷是桃花的海洋，而"丫吉宿集"就是你穿越万亩花海后停靠的港湾。相信在这个"世外桃源"，你会遇到属于自己的"十里桃花"。

Vision

The concept of "village B&B Group" started the era of home stay cluster in China, while "Yagi village B&B Group" is not only a place to live, but also integrating local culture and paying attention to the development of cultural innovation. We have specially customized display racks to show local handicrafts, hoping that visitors can also take their own memories back home at the end of the trip.

Pinggu is the sea of peach blossoms, and "YaJi village B&B Group" is the harbor where you stop to cross the sea of flowers. We believe that in this "paradise" you will meet your own "ten miles of peach blossom".

天然礼遇
邂逅丰收

北京·平谷

职住空间
Working and Living Space

东四·共享际	职住一体
南阳·共享际	戏剧工坊
星牌·共享际	职住一体
共享际 @ 国贸	职住一体
郭公庄·共享际	职住一体

Office, Apartment, Commercial and Culture Space
Space and Content

The working and living space is not a superficial meaning but can convey the core of the space. On the surface, "working" is the space of office, workshop, studio and other work scenes, while "living" is the space to solve the living needs. In simple terms is that office people also live here. However, in fact, the work-living space defines a new lifestyle with community stickiness, which solves the problem of people's stay and consumption transformation. Business forms, interactive forms and vertical communities included and extended by "working" are all content driven by innovative lifestyles. "living" is a means to enhance experience and interaction, and a space bearing for people to stay.

Operation

With the service, support and extension of "working" as the core, we will create more content, form a diverse and converged lifestyle community, and maximize the commercial value beyond job and residence.

办公、公寓、商业与文化建筑融合

关于空间与内容

职住空间并非浅显的表面含义，而是可以传达空间核心。表面上"职"是办公、工坊、工作室等工作场景的空间，"住"是解决居住需求的空间，如果简单地理解就是办公的人也在这里居住。但事实上职住空间所定义的是一种新的具有社群黏性的生活方式，解决的问题是人的停留与消费转化。"职"所包含和能够延伸出的商业业态、互动形式、垂直社群都是创新生活方式的内容驱动，"住"是增强体验和交互的手段，是让人停留的空间承载。

关于运营

以"职"的服务、扶植、延伸挖掘为核心，创造更多内容，形成多元聚合的生活方式社区，实现职住之外的商业价值最大化。

东四·共享际

项目背景

东四·共享际位于北京二环里的胡同区域，项目本身为一个废弃的酱油厂，周边被北京老城区典型的灰色坡屋顶瓦房老建筑包围，沿着一条很窄的胡同巷子，不远处是段祺瑞执政府旧址，而项目隔壁就是一些在这里居住了几十年的老居民。东四·共享际是在一个老社区里的创新的空间产品，东四·共享际的产品团队和设计建筑事务所共同研究人对空间、社群对空间的生活需求属性，来提供设计输入条件。业主运营团队和设计师的深层次全面沟通保证了产品内容和日后的互动关系与空间完美的结合。

中国·北京 东城区东四九条88号

Project Background

Dongsi 5Lmeet is located in a narrow Hutong alleyway within Second Ring Road in Beijing. The project itself was an abandoned soy sauce factory surrounded by the old gray-tiled buildings. It is not far from the historical Duan Qirui Former Government Building, and next door is the old residents living for decades. Dongsi 5Lmeet is an innovative space in an old community. Client and design firm worked together to study the social and spatial requirements of activities space to provide design input conditions. Operating team and designers in-depth communication ensured that the project content and future interaction with the perfect combination of space.

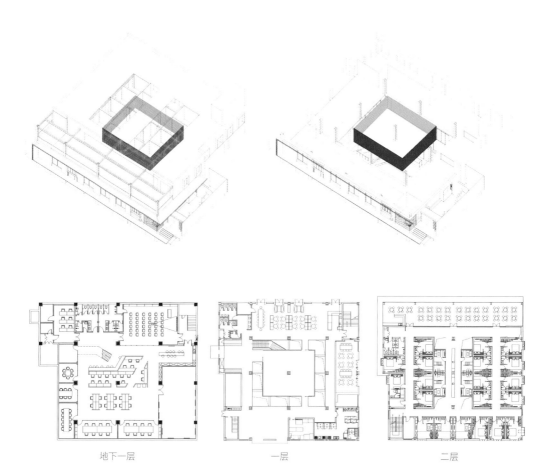

地下一层   一层   二层

设计理念

项目的设计理念来源于四合院的院落围合文化，通过切割楼板形成的"浮游之岛"，用现代手法来阐述传统建筑文化观点。而这种内部空间切割出一个通透的中心展示空间成为支撑整个项目最大的特点，一个月一百场活动也让"浮游之岛"成为使用效率最高的区域。

改造之前的建筑由于每层都是封闭的，内部光线不理想，并且视觉效果比较单调。所以在改造设计中首先通过切除每层之间的楼板，让从地下一层到屋顶所有楼层都连通，从屋顶引入光线，增加了空间的层次感。同时天井又是一个很好的广告或艺术装置展示区，为整个空间增加了灵活性及创造性。

Design Concept

The design concept of the project is originated from enclosing culture of traditional Chinese courtyard. The "floating island" is formed by cutting the floor, the traditional architectural culture view is expounded with modern technique. This transparent display space in the internal space becomes the most important feature . "Floating island" becomes the most efficient use of the region with a hundred activities per month.

The project space is very cramped and dark before renovation. So firstly cut some large holes to connect different floors and allow sunlight come in from roof, make internal space more rich. At the same time the patio is also a good advertising and art installation display area, make space more flexible and creative.

胡同里的外立面在不违背胡同大环境灰色调的基础上，把很多瓦片摞起来形成局部的墙体，自然地和胡同融合为一体。

Facade design is always very challenging in Hutong alley that can not against the gray historic environment. A lot of tile pile up to form part of the wall in the entrance.

东四·共享际整体面积虽然比较小，但是功能业态比较齐全，含有餐饮、书店、超市、办公和公寓等。

Dongsi 5Lmeet includes restaurants, bookstore, self-service shop, offices and apartments although area is not large.

运营者说

黄燕琦_东四·共享际主理人

东四社区是共享际启程的地方，也承载我们所有小伙伴的记忆与期待，在五年多的运营时间里，我们经历过波折，也一次次迎来新的人和内容，当我们所有人提到东四共享际的时候，都觉得今天的面貌得来不易。我们非常热爱这里，热爱这里入驻的每一个商家，热爱那些可爱的主理人，热爱那些在这里工作和生活、留下故事和回忆的形形色色的人们。它身在胡同，也真正把胡同里的"大杂院"文化带到了年轻人的生活之中，在这里所有的人都相亲相爱，"一句话的事儿"——这样的温暖是很难在一个现代商业体中见到的。

南阳·共享际

项目背景

在老北京的胡同中，曾经弥漫着烟火气息与艺术，形成了一种独具魅力的生活氛围。一些小剧场散落在胡同中，汇集了当时最有实力的优秀艺术家们，给当时的人们带去了一场场艺术上的饕餮盛宴，也推动了我国戏剧、艺术的发展，促进了文化交流。南阳剧场历经四十余年的历史变迁，往日的繁华已不再，亟须改造。大观建筑有幸可以参与这座曾经辉煌一时的老剧场改造。

公元前4世纪，亚里士多德表述了对戏剧本质的认识，"戏剧就是模仿"，将此概念延伸，戏剧与空间在某种程度上也具有一定的联系。空间往往也具有与戏剧一样的模仿、情景、重叠、冲突等元素。尤其是在改造项目中，空间被叠加了时间这个维度，使得历史与现在的冲突、场地与功能的冲突更加凸显。正如在一部戏剧中，不可能有永恒的平静也不可能有持久的冲突，一切起起落落的情景都有解决方法。老建筑改造中的这些矛盾与冲突，可以顺势让其表现、彰显冲突，也可以借力消减，使其和谐，或者就将不同的元素拼贴，简单坦诚地展示。

中国·北京 东城区南阳胡同6号

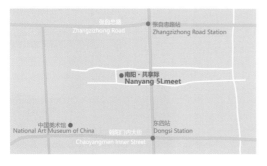

Project Background

In the Hutongs of old Beijing, the feeling of daily lives and art was once permeated, forming a unique and charming living atmosphere. Some small theaters were scattered in the Hutongs, bringing together the most powerful and outstanding artists of the time, bringing a gluttonous feast on art to the people at that time, also promoting the development of drama and art in our country, and promoting cultural exchanges. After more than 40 years of historical changes, Nanyang Theatre is no longer prosperous in the past and is in urgent need of renovation. DAGA Architects is fortunate to be able to participate in the renovation of this once glorious old theater.

In the 4th century BC, Aristotle expressed his understanding of the nature of drama. "Drama is imitation", With the extension of this concept, there is also a certain relationship between drama and space. Space often has the same elements as drama, such as imitation, scene, overlap, conflict, and so on. Especially in the renovation project, space is superimposed on the dimension of time, which makes the conflict between history and present, the conflict between site and function more prominent. Just as in a play, there can be no eternal peace or lasting conflict, so there is a solution to all the problems. These contradictions and conflicts in the renovation of old buildings can take advantage of the opportunity to show their conflicts, and they can be reduced to make them harmonious, or different elements can be collaged to show them simply and frankly.

顺势而为

项目位于北京东城区南阳胡同 6 号，属于东城区的建筑风貌保护区。受共享际的委托，大观将原为汇报演出专用剧场改造为北京城一个以戏剧演出、咖啡休闲、办公、工坊、公寓为主的多功能复合空间。保留原有的艺术氛围与历史沉淀所留下的痕迹，并给空间注入新的能量，拼贴出新与旧的结合，再创空间的辉煌，重现老剧院的艺术本色。

建筑外立面在材质上使用青砖，顺延整个胡同古朴的气质，主入口处采用柔和的弧线处理形式，形成一个具有进深感的入口等候区，同时又兼顾引导客流的作用。为了更加凸显入口的弧度，我们使用了新材料——玻璃砖，给原本古朴的青石砖添加了几笔通透的色彩，在夜晚能够给昏暗的胡同点起几盏明亮的灯。

穿过入口门厅进入大堂，临近大堂咖啡区的走廊由四段依次递减的木饰面套口分割而成，灵感来源于戏剧中的"小孔成像"原理。在走廊里，行人每靠近一步就会呈现出不同的视觉氛围。保留原有粗犷的水泥立柱并且结合新的水磨石地砖、水磨石墙裙、墙面艺术涂料及金色金属卷帘使得整个走廊充满故事性。我们顺势延伸了原本剧院的氛围，也让人们能够体验到剧场的时代交替与空间趣味性。

Follow the Nature

The project is located at No. 6 Nanyang Hutong, Dongcheng District, Beijing, which belongs to the architectural style protection area of Dongcheng District. Commissioned by 5Lmeet, DAGA Architects transformed the original theater into a multi-functional complex space in Beijing, mainly for theater performances, coffee leisure, offices, workshops, and apartments. Retain the traces left by the original artistic atmosphere and historical precipitation, inject new energy into the space, collage the combination of the new and the old, and then create the brilliance of the space and recreate the artistic nature of the old theatre.

The façade of the building uses grey bricks to extend the quaint temperament of the entire Hutong. The main entrance uses a soft arc shape to form an entrance waiting area with a sense of depth. In order to highlight the arc of the entrance, we used a new material, glass bricks, which added a few transparent colors to the original grey bricks, which can light up the dim Hutong at night.

Passing through the entrance hall to enter the lobby, the corridor adjacent to the lobby coffee area is divided by four successively decreasing wood veneer sets, inspired by the principle of "small hole imaging" in the drama. In the corridor, pedestrians will feel a different visual atmosphere every step they approach. Retaining the original rough concrete columns and combining with new terrazzo floor tiles, terrazzo wall skirts, wall art painting, and golden metal shutters make the entire corridor full of story. We extended the atmosphere of the original theater and allowed people to experience the alternation of the times and the interesting space of the theater.

野蛮生长

南阳剧场在改造前，就已经呈现出毛坯的状态，除了演出厅以外的其他空间，几乎都是粗野的水泥柱与剥落的墙皮。当我们进入场地的时候，建筑的残败在阳光的照耀下，光影的变幻之间，竟然消除了原本该有的落魄感，反而给人以一种粗野的美感。也许这就是一种空间的自我"野蛮生长"，新与旧本就不是"美"与"丑"的评判指标。所以在改造中，我们刻意保留了建筑中某些粗野的部分，让它尽情展现那种粗犷的魅力，让历史感外露，提升建筑空间结构的力量感，也让空间氛围更具有戏剧的冲突性。

建筑共有 4 层，楼梯间是建筑中最重要的竖向交通空间。楼梯间墙面一半保留原有红砖材质一半采用做旧艺术涂料材质，一半是历史一半是现在，楼梯间就像一个能够穿越时空的梦幻空间。楼梯踏步延续原本的水磨石地面搭配内嵌亚克力灯带，贯穿于整个 4 层的建筑空间，形成一个富有历史年代感的楼梯间区域。

Natural Growth

Before the renovation, Nanyang Theater was already in a rough state. Except for the performance hall, almost all the spaces were rough cement columns and peeling walls. When we entered the site, the destruction of the building was shining in the sun, and the changes of light and shadow unexpectedly eliminated the original sense of frustration and gave people a rough sense of beauty. Perhaps this is a kind of self-"natural growth" of space and the new and the old are not the criterion of "beauty" and "ugliness". Therefore, in the renovation, we deliberately retained some rough parts of the building, let it show the rugged charm, let the sense of history be exposed, enhance the sense of the strength of the building space structure, and make the space atmosphere more dramatic conflict.

The building has four floors, and the staircase is the most important vertical traffic in the building. The staircase wall half retains the original red brick material, half uses the art paint material, half is history and half is new. The staircase is like a dream space that can travel through time and space. The stair steps continue the original terrazzo floor with embedded acrylic light strips, running through the entire four-story building space, forming an area full of history.

历史的碰撞

建筑的首层功能为：接待、剧场、咖啡休闲、化妆间。踏进入口大门就来到了一个充满历史怀旧感的接待前厅与公共空间。将原本的柱子与顶棚暴露，地面用水磨石与水泥形成对比。部分空间将老剧场保留下来，老物件重新利用，也更加体现出了新旧材料之间的传承与联系。

The Collision of History

The functions of the first floor of the building are reception, theater, coffee leisure, dressing room. Stepping through the entrance door, you will come to a reception lobby and public space full of historical nostalgia. The original column is exposed, and the ground is contrasted with terrazzo and cement. In a part of the space, the old theater is retained and the old objects are reused, which further reflects the inheritance and connection between the new and old materials.

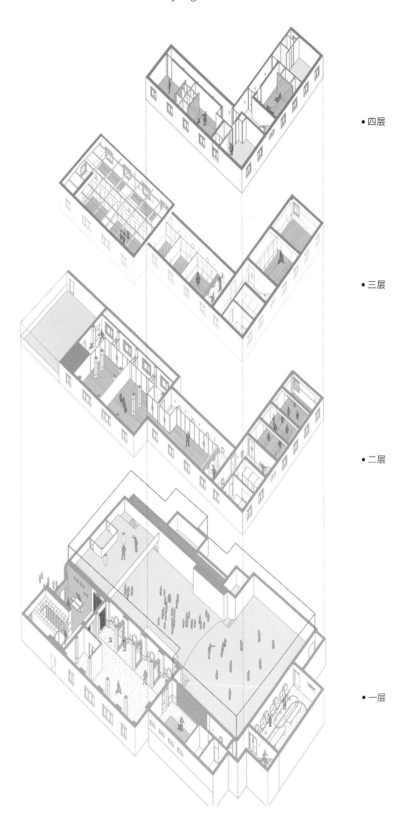

• 四层

• 三层

• 二层

• 一层

除了建筑本身极具特色之外，我们在现场还发现了另外一个特点，那就是剧院传承下来的各种老家具与物件。装道具的木头箱子、歌舞时用的大鼓、剧场的红色座椅以及传统的红绒布帘，这些物件一下子就把我们带回了那个年代。每一个物件都像是一把开启时空隧道的"钥匙"，所以在设计中，特意保留了这些物件与元素，让来到这里的人们都能回忆起当年的美好时光。

工坊L2005-L2002
Souvenir Shop

茶水间
Pantry Room

卫生间
Restroom

联合办公L2001
Co-working

In addition to the unique features of the building itself, we also found another feature at the site, that is, all kinds of old furniture and objects are passed down from the theatre. Wooden boxes for props, big drums for singing and dancing, red seats in the theater, and traditional red velvet curtains take us back to that time at once. Every object is like a "key" to open a space-time tunnel, so these objects and elements are specially preserved in the design so that people who come here can recall the good old days.

四层平面图
4 楼梯间
5 走廊
7 卫生间
17 客房1(黑白灰系)
18 客房 2（木系）

三层平面图
4 楼梯间
5 走廊
7 卫生间
14 办公区
15 茶水间
16 手工制作坊

二层平面图
4 楼梯间
5 走廊
7 卫生间
9 排练厅
13 露台
14 办公区
15 茶水间

一层平面图
1 门厅
2 保安室
3 休闲区
4 楼梯间
5 走廊
6 剧场道具通道
7 卫生间
8 化妆间
9 排练厅
10 演播厅
11 设备间
12 疏散通道

南阳·共享际的四层的功能是以公寓为主，在这样的一个空间内，可以满足人们工作、休闲、居住、生活、文娱等各方面的需求。

以黑白灰为主色调的公寓，玄关、卫生间与卧室区域形成灰、黑两个独立的色块，简洁干练。卧室大胆地使用水泥墙面漆，搭配富有戏剧感的软装家具，创造一种开放而充满层次的环境。

木色系的客房中，卧室背景墙采用胡桃木色的木格栅与大块饰面板结合的方式，同时兼顾床头板的功能。做旧金属色壁灯与复古开关面板的搭配让空间的细节处更加突显精致。卫生间与衣帽间的开敞式运用，舒适灵活，让租客可以在细节之处体验到空间中蕴含的剧院元素。

The fourth floor of 5Lmeet is mainly composed of apartments. In a space like 5Lmeet, it can meet people's needs for work, leisure, living, and entertainment.

The apartment with black, white, and gray as the main color, the entrance, bathroom, and bedroom area form two independent color blocks of gray and black, which are simple and capable. The bedroom uses cement wall paint with dramatic furniture to create an open and hierarchical environment.

In the wood color apartment, the background wall of the bedroom adopts the combination of walnut wood panel, taking into account the function of the headboard at the same time. The combination of distressed metal wall lamps and retro switch panels makes the details of the space more exquisite. The open use of the bathroom and cloakroom is comfortable and flexible so that tenants can experience the theater elements contained in the space in detail.

主理人说

赵禹清_共享际品牌总监/南阳剧场主理人

戏剧，在今天过时了吗?在同一个空间，演员和观众共同分享同一个故事和情感，直达彼此体验，悲欢相通。

南阳共享际前身是中国演出集团的排练场地，改造之初，我们设想作为戏剧爱好者，以客人的身份抵达这个空间，期待将有什么样的体验。以至于在项目正式进入运营阶段时，由于不想错过每位客人进入或离开项目时的第一感官反馈，我便将前台当作了自己的工位。现在的南阳共享际是一个极具戏剧沉浸感的场域，剧场、工坊甚至是公寓都是戏剧的舞台，我们希望通过空间去表达对生活与文化的理解，它的职能不仅是为了满足和服务社区公众，也是文化和资源得以持续的容器。

安妮_"安妮看戏"主理人/南阳剧场戏剧顾问

莎士比亚说"整个世界是一座剧场"，这句话让无数热爱戏剧的人怦然心动。共享际从北京南阳胡同出发，把戏剧装进多元丰富的城市空间，描点连线，未来大约是星罗棋布的城市图景。在这个过程中，社区与现场艺术自然联结，交流、构建、生长，浪漫寓言之外，世界真的成为一座剧场。

翔子_南阳剧场戏剧主理人

在南阳共享际的运营过程中，团队一直秉持着树立社区共同价值观，激发社区自我成长的运营理念。在一个社区中，人之间的关系会相对紧密，人之间建立的联系也是多维度的。社区最大的公约数是一个共同价值观，认同这个价值观，社员就会更支持运营者的工作，也就是所谓的劲往一处使。使得社员成为运营的主导，自发配合创造社区内容更新，实现社区的自我成长。

星牌・共享际

项目背景

星牌・共享际位于北京市大兴区兴业大街三段，临近黄村火车站。项目规模约 8600 平方米，整体建筑共六层，包含了地上五层与地下一层，改造前为一处空置的社区配套商业。"共享"是这个时代孕育出的具有代表性的一个关键词，从 co-working 到 co-living，慢慢成为一种最新的生活方式的倡导，新的生活方式带来的强大能量已经冲破了传统的商业模式。此次项目是针对陈旧的社区配套商业的改造，通过改造将空间打造成为集办公、公寓、商业等业态于一体的一站式 co-living 共享生活社区，创造更丰富的共享生活体验场景。

中国・北京 大兴区兴业大街

Project Background

Xingpai 5Lmeet is located in the third section of Xingye Street, Daxing District, near Huangcun Railway Station. The area is 8600 square meters, and the whole building consists of six floors, with five floors aboveground and one floor underground. It was a vacant community supporting commercial before the renovation. "Sharing" is a representative keyword that was born in this era. From the concept of co-working to co-living, it gradually becomes an advocate of the latest lifestyle. The powerful energy brought by the new lifestyle has broken through the traditional business model. The project is designed to transform the old community supporting commercial into one-stop co-living community, which is a combination of business formats include office, apartment and commercial to create a richer shared life experience scene.

城市语境

在此次项目的设计过程中，设计师以 co-living 的生活方式为灵感打造了一个新的语汇——"共享站"，尝试在两种功能空间的交界处设置一种新的混搭空间，这个空间在功能上可以同时满足两个主功能区所需要的服务社交功能；在空间上成为整体设计上的重点；同时又解决了空间效率的问题。

Urban Context

Regarding the design concept of this project, designers are inspired by the shared lifestyle and create a new vocabulary-"communion station". They attempt to set up a new complex space at the junction of the two functional spaces, and this space can simultaneously satisfy the social functions of the services required by the two main functional areas. The new space not only becomes the focus of the design, but gives the solution of space efficiency.

设计理念

共享站以独立空间呈现，在每一层将办公、公寓、商业等业态的交界处做了富于弹性的设计处理，地下一层和一层局部打通，打造了一个综合性活动空间，满足路演、讲座及各种社会活动场地使用功能，是一个灵活的公共性活动空间。三层的共享站具有接待水吧以及小影院的附加功能区域，四层的共享站具有超市以及中式厨房的附加功能区域，五层的共享站具有健身房、洗衣房以及厨房的附加功能区域。共享站使空间的功能性更加丰富，不同区域里的人可以有更多机会交流互动，真正体会到共享的乐趣。

空间中的设计语言以彩色的线性元素贯穿始终，满足功能性的同时达到视觉统一的效果。简化设计使整个空间仿佛形成一个白色画布，再配以明快的彩色线条，从隔断到墙面图案，如同跳动的音符跃然纸上，增强了空间的立体感和活力氛围,元素的重复使用打造了具有记忆点和活力氛围的共享空间环境。

Design Concept

The communion station is presented in an independent space. The designers utilize the flexible design at the junction of office, apartment and business at each level. The spaces of the ground floor and the first floor are connected to create a flexible public event space that meets the functions of roadshows, lectures, and various social events.

The third floor includes reception area, bar area and small theater. The fourth floor includes supermarket and Chinese kitchen, and the fifth floor includes gym, laundry and kitchen. The design of the communion station makes the space level more diverse and functional, and more importantly, it provides more opportunities for people to exchange and interact with each other and thus experience the enjoyment of sharing lifestyle.

The design language uses color linear elements throughout the space to satisfy the functionality as well as achieving a uniform visual effect. The space has been simplified designed as a white canvas. From partitions to wall patterns, these bright colored lines are like beating notes on the paper, enhancing the three-dimensional visual effect and vibrant atmosphere of the space. The design elements are used repeatedly to create a unique co-living space environment with vivid memory.

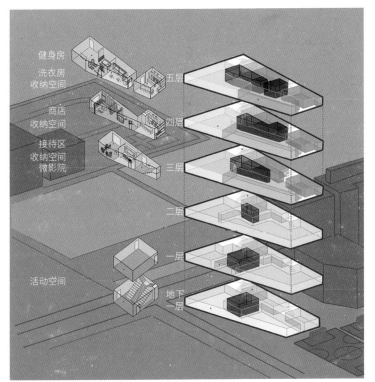

项目信息

项目名称：星牌·共享际

地点：北京市大兴区兴业大街

内容策划：优享与行策划咨询有限公司

甲方团队：王翔，吴雪

设计公司：GB SPACE 高白空间设计事务所

设计师：高文毓，白杨，张婧，廖青

竣工时间：2018 年 12 月

项目面积：8600 平方米

共享际 @ 国贸

项目背景

共享际 @ 国贸项目紧邻三环，位于 CBD 核心区内，与国贸三期隔街相望，地理位置绝佳，但超短租期，老建筑微改造，如何运营创造高租金收益，是非常难以平衡投入与收益的项目。在经过多轮策划与设计后，项目整体定位为：CBD 一站式共享社区，内容孵化实验场。功能包括长租公寓、酒店、商务会议、美食孵化器、运动健身、联合办公等，为 CBD 青年白领提供多元的生活方式聚合空间。

中国·北京 朝阳区CBD

Project Background

5Lmeet@Guomao is close to the Third Ring Road, located in the core area of CBD, and looks across the street from International Trade Phase III. It has an excellent location, but it is very difficult to balance investment and income for the project with ultra-short lease period, old building micro transformation, and how to create high rental income through operation. After several rounds of planning and design, the overall positioning of the project is as follows: CBD one-stop co-living community and content incubation experiment field. The functions include long-rent apartment, hotel, business meeting, food incubator, sports and fitness, co-working, etc., providing diversified lifestyle gathering space for young white-collar workers in CBD.

设计理念

运营者说

王翔_共享际设计总监/首席建筑师

国贸，这样一个城市高端商务的聚集区，共享际却带给了它一个充满活力的年轻社区，一栋经营期极短的建筑，对于运营和改造的挑战不仅仅是数据层面，更大的困难是可落地的内容用何种经营模式去对冲风险，因此我们在设计和改造中将固定投入压缩到极限，运营中将内容孵化作为投入的重点，实现了与一个个强力品牌的合作和对每一点空间的合理开发。每一份投入都成长为自身品牌的矩阵扩张，成为真正的商业孵化器。

项目信息

项目名称：共享际 @ 国贸

地点：北京市朝阳区建国门外大街 1 号

内容策划：优享与行策划咨询有限公司

甲方团队：钟宇辰

设计公司：大观 DAGA 建筑设计有限公司

设计师：申江海

竣工时间：2018 年 12 月

项目面积：18000 平方米

本项目未对建筑结构改动，主要改造内容为首层商业与大堂空间，客房中改造均为软装提升，介于本项目特殊性，进行适度改造实现项目经营与建筑物业的保值增值

郭公庄 · 共享际

项目背景

郭公庄 · 共享际地处北京市丰台区西南四环外丰科路东侧，北京方向小区内配建集中商业体，全楼为商业功能，定位为社区配套型商业中心。项目原始建筑面积约 50000 平方米，本次空间改造为项目一部分，建筑面积 16000 平方米，一层二层为公交场站停车场，三层至五层为商业功能，从地面有专门的出入口和垂直交通到达三层及以上楼层。项目紧邻北京方向小区，距离永旺梦乐城不足 2 千米，距离世界公园约 5 千米，周边密布万科 · 蓝山、中海、万科 · 西华府等一系列中高端居住社区，是典型的社区商业改造项目。

中国 · 北京 丰台区郭公庄

Project Background

Guogongzhuang 5Lmeet is located on the east side of Fengke Road outside the Fourth Ring Road in the southwest of Fengtai District. It was built as a centralized commercial building for the direction Beijing community. The whole building is for commercial functions, and it is positioned as a supporting commercial center for the community. The original construction area of the project is about 50000 square meters. This space transformed is a part of the project, with an area of 16000 square meters. The first and second floors are for bus station and parking, the third to the fifth floors are for commercial functions, and there are specified entrances and vertical traffic from the ground to the third and above floors. The project is close to the Beijing direction community, less than 2 kilometers away from Aeon Mengle City, about 5 kilometers away from the World Park, surrounded by Vanke · Lanshan, Zhonghai, Vanke · Xihuafu and a series of medium and high-end residential communities, which is a typical community commercial transformation project.

城市语境

原本封闭的建筑立面将建筑塑造成一个独立的与外界没有交流机会的盒子空间，在开放、共享、交流的理念下，设计将公寓空间的顶部及各层楼板打开，引入阳光，创造一个舒适自然的共享空间。利用建筑三层顶部的局部露台，为商业、办公、公寓功能创造交汇的开放空间，向内可以实现功能业态的互动与交流，向外可以由此远眺城市，创造一个链接城市的窗口。

Urban Context

The originally closed facade of the building shapes the building into an independent box space with no opportunity to communicate with the outside world. Under the concept of openness, sharing and communication, the design opens the top of the apartment and every floor to bring in sunlight and create a comfortable and natural sharing space. The partial terrace on the third floor creates an open space for the intersection of commercial, office and apartment functions. The interaction and communication of functional formats can be realized inwards, and the city can be overlooked outwards, creating a window linking the city.

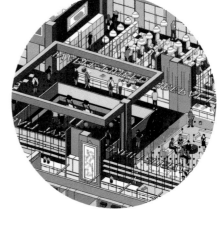
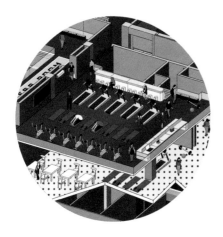
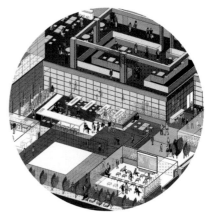

Design Concept

The commercial atmosphere around Guogongzhuang Subway station is not conducive to the development of community-oriented commerce. There are two large shopping centers, Aeon mengle City and Fengtai Wanda Plaza, within 2 kilometers. Besides, the huge office and residential population of the headquarters base forms a strong regional siphon effect. How to get different flow within the scope of large shopping center 2 kilometers is an important goal of this project, so after investigating industrial structure and the surrounding residential population age structure of headquarters office , the whole project may be defined as sets of office, residential, innovation business in the sharing paradise, attracting peripheral industry upstream and downstream industry in office. At the same time, IP network celebrity business is introduced for incubation and cultivation, and convenient business services are provided for surrounding residents within 1 kilometer.

Each function locates in the three-floors building in vertical distribution, each floor layout of three functions, to ensure the intersection of human flow moving lines, which create new chemical reactions. New business community, and undeveloped area lacks the steady stream of people and commercial production base, the design need to solve not only the pattern of space and decoration,but more thinking how good combination with the content, the space for commercial facilities from customers and achieve retained.Let the office efficiency, make live comfortable, make comerce convenient, and is equipped with unicom buffer between each function area. Let the communication happen, let the interest between people become the characteristics of the space. So that guests feel from the heart of sharing the beauty of life.

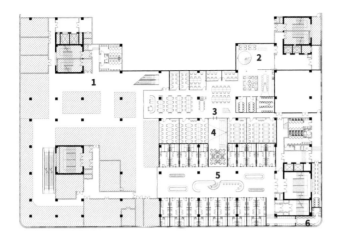

1 商业区
2 办公前厅
3 联合办公区
4 多功能厅
5 公寓客厅
6 洗衣房

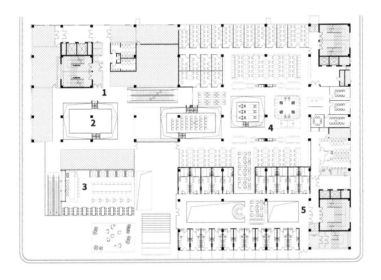

1 商业区
2 多功能厅
3 户外露台
4 联合办公区
5 公寓

设计理念

郭公庄地铁站周边商业氛围并不利于社区型商业的开展，2千米范围内有永旺·梦乐城、丰台万达广场这两个大型购物中心，且围绕总部基地庞大的办公和居住人群，形成强大的区域虹吸效应，而本项目周边则主要为居住社区，如何在大型购物中心2千米范围内实现社区商业的差异化引流是本项目实现的重要目标，因此在考察总部办公产业结构与周边居住人群年龄结构后，将整个项目定位为集办公、居住、创新商业于一体的共享生活乐园，吸引周边产业上下游行业入驻办公，为办公人群提供居住配套，同时引入IP网红商业进行孵化培育，为项目人群与周边居民提供1千米内的便利商业服务。

整个项目将各个功能垂直分布在三层建筑内，各个楼层都布局三个功能，能够确保人流动线的交汇，产生新的化学反应。新开发、待发展区域的社区商业缺少稳定人流和商产基础，设计需要解决的不仅仅是空间的格局和装饰，更应从客群的引入并实现留存出发，思考如何将空间与内容进行良好的结合，让商业便利，让办公高效，让居住舒适，同时又在各功能间设有连通缓冲的区域，让交流发生，让人与人之间的趣味成为空间的特质，从而让客人发自内心地感受共享生活的美好。

运营者说

王翔_共享际设计总监/首席建筑师

对于城市近郊上万平方米的购物中心改造与运营，郭公庄·共享际是一次大胆的尝试，在周边配套和客流尚需培育的城市空间，将自身打造成为消费与生活的目的地，将居住、办公、商业空间紧密地融合，创造一个向内生长，向外延伸的独特空间。为了创造场景的多变与融合，将三种业态在每一层都均匀布局，虽然这增加了运营的难度，但无形中创造了一个社区中的社区，不断强化生活的归属感，为运营而设计，为生活而设计。

项目信息

项目名称：郭公庄·共享际
地点：北京市丰台区郭公庄中街
内容策划：优享与行策划咨询有限公司
甲方团队：王翔，钟宇辰
设计公司：大观 DAGA 建筑设计有限公司
设计师：申江海，任晓伟
竣工时间：2018 年 1 月
项目面积：16000 平方米

办公空间

Working Space

昆明大悦城 · 优客工场	共享办公
鲜鱼口宝马 X2 · 共享际	共享办公
金茂资本 J SPACE	定制办公
华贸中心 · 优客工场	定制办公
兆维 · 优客工场	共享办公
南京高铁 · 优客工场	共享办公
阳光 100 · 优客工场	共享办公

Co-working and Customized Space
Space

After years of development, the co-working space has gradually matured and expanded its business services to finance, tax, legal and business-oriented services, bringing convenient services to the settled enterprises and paying more attention to the support and assistance at the business level. In this way, we can make use of the huge membership system and industry field to create a closer enterprise and content matrix, create more cooperation, and create a better environment for the growth and development of member enterprises.

Operation

The co-working scene is no longer a simple business of space and desk, but a business scene of zero distance contact between enterprises and members through offline office scene. It becomes a portal platform enterprise for commercial traffic, aggregating industries and content, and upgrading services into strategies.

共享办公、定制办公

关于空间

联合办公空间经历多年发展，已逐渐成熟，将企业服务的内容扩展为集财务、税务、法务、业务为主的服务型平台，为入驻企业带来便捷服务的同时，更加注重企业业务层面的扶植与协助，从而利用庞大的会员体系与行业领域，创造更加紧密的企业和内容矩阵，创造更多合作，为会员企业创造更好的成长与发展环境。

关于运营

联合办公的场景已经不再是简单的空间与办公桌的业务，而是通过线下办公场景创造企业与会员零距离接触的商务场景，成为商业流量的入口型平台企业，聚合产业与内容，让服务升级为战略。

昆明大悦城·优客工场

项目背景

昆明大悦城·优客工场是大观在昆明做的第二个项目，地处昆明千年商脉——老螺狮湾，依昆明母亲河盘龙江畔而建造，占据昆明市一环及多条城市主干道。优客工场位于大悦城的11、12层空间。由大观与优客工场团队紧密合作，通过分析昆明当地特色，探索和营造符合现代文化环境和办公方式的多元工作空间。

中国·昆明
西山区环城南路大悦城购物中心6栋11、12层

Project Background

Kunming Joy-City Ucommune is the second project done by DAGA Architects in Kunming. It is located in Kunming's traditional commercial district-Laoluoshi Bay and is built on the bank of the Kunming's mother river Panlong River, occupying one or more sections of Kunming City. Ucommune is located on the eleventh and twelfth floors of Joy City. DAGA works closely with the Ucommune team to explore and create a diversified workspace in line with the modern cultural environment and office style by analyzing the local characteristics of Kunming.

地域特质

在昆明乃至整个云南地区，都有着很鲜明的民族、地域、文化特征，而且都保存得比较完好。云南之所以保留着独特的地域氛围，正是因为它没有在城市化的浪潮中被时代背景所吞没。在城市中，从来不缺少现代化的建筑，但是在千篇一律的发展过程中，都在逐渐地趋同，城市空间需要体现地域性，不仅仅是对传统、文化的一种保护，也是一种城市发展多元化的体现。

云南特有的自然条件，在这里形成了许多极具特色的梯田与竹楼，附近的村落分布在梯田四周，形成一种聚合式的布局。我们将这些元素与联合办公中现代社区的功能结合，形成了一个围绕着中间共享空间布局，四周有各种功能环绕的基于地域性的空间布局模式。同时，联合办公的功能也营造了一种社区化的办公模式，打破了传统写字楼对人造成的消极固化的影响。

Regional Characteristics

In Kunming and even the whole Yunnan region, there are very distinct ethnic, regional and cultural characteristics, and they are all well preserved. The reason why Yunnan retains its unique regional atmosphere is that it is not been changed by the background of urbanization. In the city, there is never a lack of modern buildings, but in the process of uniform development, they are gradually converging. Urban space needs to reflect the regionality, not only a kind of protection of tradition and culture but also an embodiment of the diversification of urban development.

The unique natural conditions of Yunnan have formed many unique terraces and bamboo buildings here, and nearby villages are distributed around the terraces, forming an aggregated layout. We combine these elements with the functions of the modern community in the co-working office to form a regional-based spatial layout pattern around the middle shared space and surrounded by various functions. At the same time, the function of co-working also creates a community-based office mode, breaking the negative impact of traditional office buildings on people.

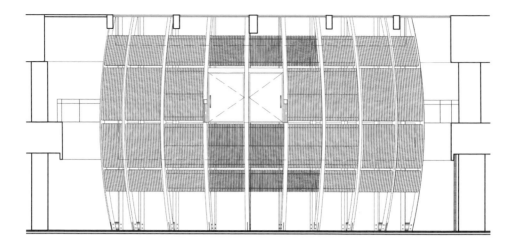

空中竹楼

项目位于整幢大楼的最顶层，我们将这两层以及屋顶作为一个整体来设计，并且在最中心植入一个空中的竹楼，在建筑中营造出一个巨大的围合空间，消解了建筑中矩形柱网的均质感。竹楼是云南地区傣家的标志民居，我们将此元素在公共区域延伸，结合地域性的材料，让整个设计能够为人们传达出城市的个性与特征。

云南省的天然竹林面积有 380 万亩，竹林类型共有 30 多种，天然竹林类型及面积居全国第一位，是我国乃至世界公认的竹类来源大省。竹子作为一种建筑材料，已经在许多地域建筑中运用，竹材质的柔性与刚性性质也给空间提供了更多的可能性。

Bamboo Tower in the Sky

The project is located on the top floor of the whole building. We designed these two floors and the roof as a whole, and planted a sky bamboo building in the center, creating a huge enclosed space and eliminating the uniform texture of the rectangular column net in the building. The bamboo tower is a symbolic residential house of the Dai family in Yunnan. We extended this element in the public area, combined with regional materials, so that the entire design can convey the personality and characteristics of the city to the people.

The area of natural bamboo forest in Yunnan Province is 3.8 million mu, and there are more than 30 types of bamboo forest. The type and area of natural bamboo forest ranks first in the country, and it is recognized as a major bamboo source province in China and even in the world. As a kind of building material, bamboo has been used in many regional buildings, and the flexible and rigid nature of bamboo material also provides more possibilities for space.

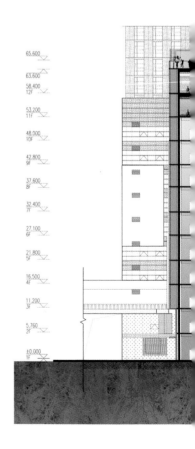

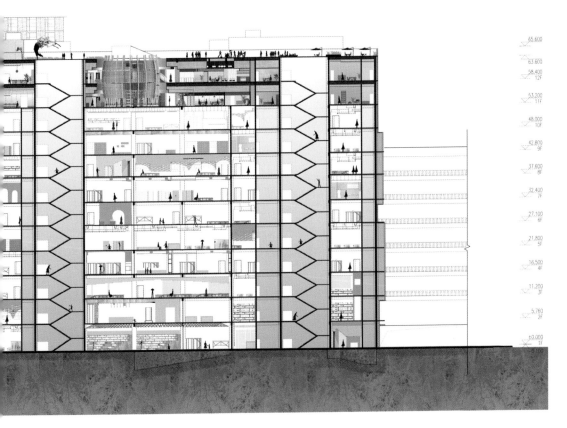

空间形态

空间形态来源于云南本地竹楼结构及梯田形态，通过竹制材料的介入，融合云南传统竹编工艺，强化本地特色。竹材的柔性，使得建筑内部空间的灵活性更强，整个色彩也使得室内更加柔和、舒适，营造了良好的办公与休闲环境。

空间布局围绕中庭路演区展开，将共享功能模块布置在建筑几何中心使得空间配比相对均衡。通过增设连通各层和屋顶花园的楼梯，打破各层界限，增强空间的关联性。采光天窗增加了阳光的射入，使得进深过大的室内每个角落都可以有自然光，给办公者营造了良好的氛围。

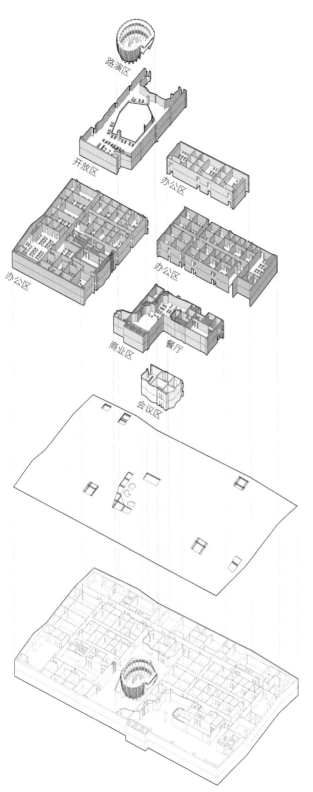

路演区

开放区

办公区

办公区

办公区

商业区

餐厅

会议区

Spatial Form

The spatial form comes from the local bamboo building structure and terrace field form in Yunnan, which strengthens the local characteristics through the intervention of bamboo materials and the integration of Yunnan traditional bamboo weaving technology. The flexibility of bamboo makes the interior space of the building more flexible, and the whole color also makes the interior softer and comfortable, creating a comfortable office environment.

The spatial layout is carried out around the atrium road performance area, and the shared function modules are arranged in the center of the building geometry to make the space ratio relatively balanced. Through the addition of stairs connecting each floor and the roof garden, the boundaries of each floor are broken and the relevance of the space is enhanced. The daylight skylight increases the exposure of sunlight so that there can be natural light in every corner of the room, creating a good atmosphere for office workers.

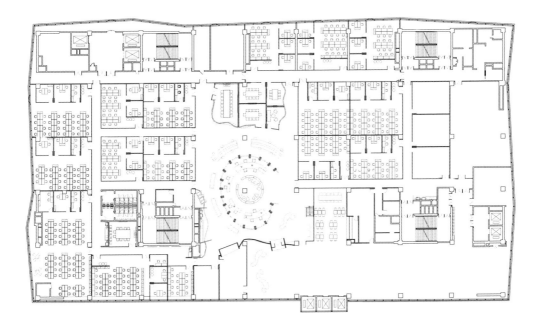

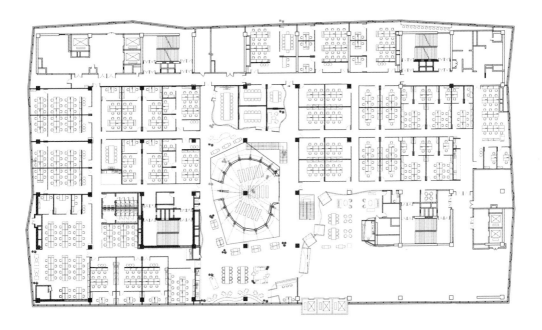

天窗细节

运营者说

关心_优客工场CEO

目前我国的城镇化率已经超过 60%。新型城镇化的下半场，如何提升城市发展质量和人民的生活品质，是新阶段我们需要探寻的新问题。2021 年，城市更新首次被写入政府工作报告，这也意味着城市更新上升为国家战略。在探究城市更新问题的过程中，共享际和优客工场在居住空间、职住空间、办公空间和新消费空间等方面积累了诸多可借鉴的案例。从本书的案例中，也许我们能更好地了解城市更新的理论脉络、目标取向、机制模式和未来趋势，帮助我们打开城市更新的新思路。

Project Background

A historic change has taken place in a building in Xianyukou Hutong, Qianmen, Beijing. Commissioned by BMW and 5Lmeet, DAGA renovated a small building in the old alley of Xianyukou. It broke away from the traditional office model, integrated garage, gas station, automobile repair shop with the same elements of metal, and created a public sharing space with BMW X2 as the theme, which is responsible for the operation of 5Lmeet. In the meantime, it also brings new vitality to this century-old commercial street.

【X2 共享际】
主题的共享办公空间由德国汽车品牌BMW与共享
它在鲜鱼口百年商业老街内与周围环境产生强
反其道而型"的精神和创新的设计理念，营造出

本着
品牌BMW X2 主色全色元素力
产品墙、博士车多能百、加油站、车库、户
空间，其概念也秉承了共享际、直
车与共享空间的跨界合作，也是一次成功的会

设计主创团队：BMW、共享际、大观

鲜鱼口宝马
X2·共享际

项目背景

北京前门鲜鱼口胡同里的一栋建筑，发生了一次
历史性的变革。受宝马和共享际委托，大观把鲜
鱼口老胡同里的一栋小楼翻新修整，打破以往传
统办公室的模式，把车库、加油站、汽车修理厂
与相同元素金属融为一体，打造了一个以宝马 X2
汽车为主题，共享际来负责运营的办公共享空
间。与此同时，也为这条百年历史的商业老街带
来了新的活力。

中国·北京 东城区鲜鱼口胡同

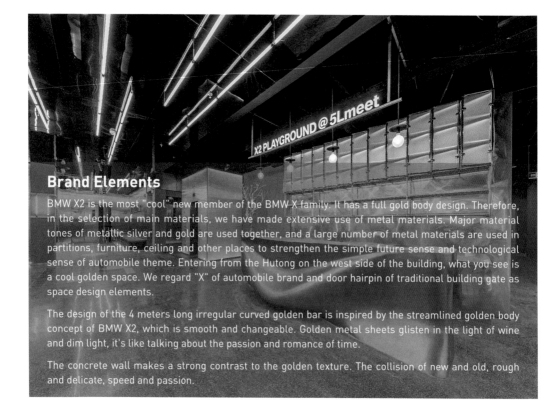

Brand Elements

BMW X2 is the most "cool" new member of the BMW X family. It has a full gold body design. Therefore, in the selection of main materials, we have made extensive use of metal materials. Major material tones of metallic silver and gold are used together, and a large number of metal materials are used in partitions, furniture, ceiling and other places to strengthen the simple future sense and technological sense of automobile theme. Entering from the Hutong on the west side of the building, what you see is a cool golden space. We regard "X" of automobile brand and door hairpin of traditional building gate as space design elements.

The design of the 4 meters long irregular curved golden bar is inspired by the streamlined golden body concept of BMW X2, which is smooth and changeable. Golden metal sheets glisten in the light of wine and dim light, it's like talking about the passion and romance of time.

The concrete wall makes a strong contrast to the golden texture. The collision of new and old, rough and delicate, speed and passion.

品牌元素

BMW X2 作为 BMW X 家族中最"炫酷"的新成员，全金色的车身设计，因此在主材料选择上，我们大面积运用了金属材质。以金属银和金黄两种色彩的主要材质基调配合使用，并在隔断、家具、顶棚等多处运用了大量金属材料，强化简约的未来感和科技感的汽车主题。从建筑西侧的胡同进入，映入眼帘的是一个炫酷的金色空间，把汽车品牌的"X"和传统建筑大门的门簪作为空间的设计元素。

4 米长不规则的曲面金色吧台，设计灵感源于 BMW X2 的流线型的金色车身概念，流畅而富有变化。金色的金属板在酒光之中，在昏黄的灯光映衬下，闪闪发光，像是诉说时光里的那些激情浪漫。

混凝土的墙面与金黄色的质感形成强烈的对比，呈现新与旧的碰撞、粗犷与细腻的碰撞、速度与激情的碰撞。

运营者说

鲜鱼口宝马 X2・共享际的设计初衷并不单纯只是在前门大街历史片区内完成一个共享办公和展览空间，更多的是希望在古朴的历史街区内创造出一个别有洞天的新场景，室内与室外的文化气质形成巨大的反差，使用最现代的材料、造型、灯光、完成从"旧"到"新"颠覆性的改变。

一个空间在不同时间分别呈现汽车发布展厅和共享办公的空间场景，在展览结束后，共享办公继承了展厅的风格与元素，又重新生长出新的功能和场景，也让共享办公的空间有了汽车展厅的基因，形成独树一帜的室内风格，是一次场景转换的成功尝试。

吴雪_共享际首席室内设计师

Design Concept

On one side of the bar is a box wrapped in gold. The bright golden curtain extends to the whole space, gleaming with golden light. The softness of the golden curtain and the reflection of the mirror stimulate people's sense of touch and sense organ. When the curtain is closed, people can hold private activities and performances; when the curtain is opened, it can be used as a gymnasium for Yoga Calisthenics. Through the translucent metal curtain, it can be merged with the outer space or separated into separate spaces.The conference room is another all golden space in the whole space. Metal plates reflect constantly each other, against the ceiling lights, there is an infinite extension of space feeling. As if we are sitting in BMW X2, and we keep moving in the direction indicated by the arrow.

BMW X2 emphasizes rebel against, breaking bondage and being different. Corresponding to this, DAGA believes that the office space in Xianyukou Hutong should also be bright, dazzling purpose, technological and minimalist. Unlike many dull and tasteless office spaces, the whole space is brightly lit. The goal is to put people in gold as soon as they enter, which breaks the traditional definition of ordinary space type, vitalizes the office environment and captures the charm of the car. This is not only the office space, but also the people's sensory laboratory.

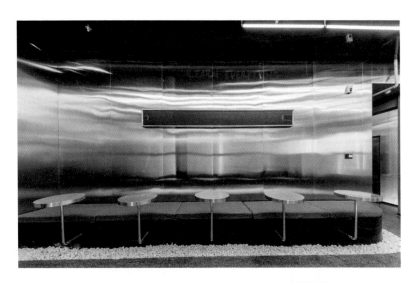

设计理念

酒吧的一侧是一个被金色包裹着的盒子，金色明亮的幕帘延伸到整个空间，闪烁着金色的光芒，这个金色幕帘的柔和与镜子的反射激发了人们的触觉和感官。幕帘拉上时，人们可以举办私密性的活动表演；打开幕帘，可以作为瑜伽、健美操的运动健身场所。通过幕帘这种半透明的金属，可以和外面的空间合并在一起，也可以分开，成为彼此独立的空间。会议室是整个空间的另一个全金色的空间，金属板之间相互不停地反射，在顶棚灯光的映衬下，有一种无限延伸的空间感受。仿佛坐在 BMW X2 里，顺着箭头指示的方向不断前行。

BMW X2 强调的是叛逆、打破束缚、与众不同。与此相呼应的，大观认为鲜鱼口的办公空间也应该是明亮的、炫目的、科技的、极简的。不同于很多办公空间的平淡无味，整个空间灯火通明，意在让人一进入，就置身于金色中，打破人们对普通空间类型的传统定义，在办公的环境中注入活力，捕捉汽车的魅力。这里不仅是可以办公的空间，更是人们的感官实验场。

金茂资本
J SPACE

项目背景

在这个"宇宙的中心"，高学历、高技术、高追求的科技精英们正在用一行行的代码影响着时代的发展，作为设计师我们无法改变"996"，但是我们可以创造一个酷炫的办公空间，让办公回归酷的属性。因此我们将科幻造园植入办公空间，营造一个前所未见的异次元办公空间，在这个空间所有的功能都与造景结合，给使用者一个场景感极强的沉浸式科幻办公体验。

我们将这个场景设定为关于平行时空的穿越故事，以此来致敬带给我们灵感的《2001：太空漫游》，以走入异次元空间的每一帧画面串连每一个重要的空间。

中国·北京 海淀区西二旗大街39号

故事线索

第一章：人类的登场 / Ascent of Man

2020 年，一个发亮的盒子穿越平行宇宙从夜空降临，消失在北京市海淀区西二旗大街 39 号楼……

In the year 2020, a luminous box descended from the night sky through a parallel universe and disappeared in Building 39, Xierqi Street, Haidian District, Beijing……

第二章：穿越星之门 / Through the Star Gate

他披着普通建筑的外衣，无意中打开他的大门，穿过微缩的黑洞隧道，就闯入了异次元空间。在这里空间似乎是无限的，一切可能会发生的事情必然会发生，就此我们开始一场异次元空间之旅。

An ordinary building, but while we opened the door by accident and entered the multidimensional space through the tunnel of the black hole. The space seems infinite, which anything can happen here.

Finally, let us begin a journey into the multidimensions.

第三章：进入眼睛

隧道之门打开，失落文明的遗迹中露出异次元之眼。你在注视着他，他也在注视着你，无限地延长，不断地掉落，穿越宇宙的中央站。

Into the Eye

When the door opens, revealing the eye of the multidimensional space from a lost civilization. The tunnel extends and falls endlessly through the center of the universe.

第四章：孤寂

1：4：9 的精准比例，可想而知在地球之外，早就有外星智慧生命，这面尘封已久的孤寂山谷和钻石切割的黄金穿梭梯坚不可摧，创造者十分强大。

Alone

The precise ratio of 1：4：9 can be imagined that there are already extraterrestrial intelligent lives outside in the universe. The creator of the structure is indeed very powerful, as both of the lonely valley and the golden stairwells are indestructible like diamonds.

第五章：真空之中

真空是我们的采光中庭，中庭的每一个楼层都可以注视孤寂山谷和黄金穿梭梯，而黄金穿梭梯在真空之中担当着运输的工作，可以到达你想去的任何地方。

In Vacuum

"The vacuum" is filled in the building atrium, you can watch over both of the lonely valley and the golden stairwells at any level. The golden stairwells acting as a transportation vehicle that can send you to any places you want to go to.

第六章：巡航模式

正式入驻此空间站，异次元接待的工作人员将带给你一个前所未有的服务感受及体验。正如智慧大脑的储藏库一般，里面内置的内存条在考量人类健康问题与空间环境的关系并更具有针对性地调整适宜人类的环境系统。

Cruise Mode

We entered into the space station, the multidimensional reception staff will bring you a service that you have never experienced before. Just like the brain of the artificial intelligence, the memory strip is built in to balance up the relationship between the lifestyle of mankind and the space environments to find the best environmental system which suits the human beings.

第七章：神奇的房间

异次元空间里神奇的房间，不断上演着关于穿越文明会议，你可以坐在原地等待什么事发生，也可以打开舱门，走到外面去向周围的现实挑战。

The Room

There are constant meetings about traveling through the civilizations where you can either watch for things to happen next or open up the hatch and meet the challenges in reality around you.

第八章："知的需求"

异次元空间的能量补给中枢。圆的造型模拟异次元之眼，在无形中给予内置储存能量。

"Need to Know"

This is the energy supply center in multidimensional spaces. The shape of the circle simulates the eye of different dimensions, imperceptibly giving it built-in energy storage.

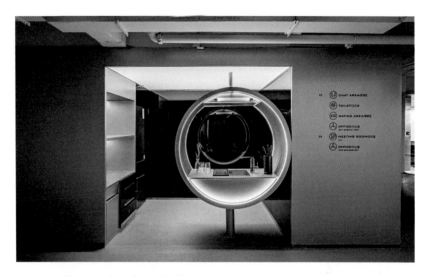

第九章：秘密

在这个异次元空间的隐蔽的角落，散布着关于穿越文明秘密的信息，破译这些信息就有可能发现异次元的秘密。

The Secret

In the hidden corner, information about the secrets of civilization was scattered everywhere. By deciphering this information, we can discover the secret in the multidimensions.

第十章：异空

脱离地心引力而漂浮存在的异次元空间悬浮山岩使人强烈地感觉到周围的景色会突然消失。

The Alien Sky

The rocks in multidimensions float free in gravity, giving a strong feeling of the surrounding landscape that will disappear in a sudden.

1 入口门洞
2 通道
3 中庭
4 水景区
5 绿植种植区
6 值班室
7 电梯
8 走道
9 电设备间

1 采光中庭
2 黄金楼梯
3 三角观景打卡区
4 洽谈区
5 黑胶等候区
6 水吧区
7 电梯观景区
8 弱电间 / 机房
9 电话间
10 洽谈间
11 会议室
12 开敞办公区
13 办公区

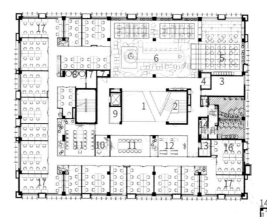

1 采光中庭
2 黄金楼梯
3 轻餐后厨区
4 星巴克自助区
5 多功能区
6 超级吧台
7 弱电间 / 机房
8 电话间
9 电梯观景区
10 洽谈间
11 会议室
12 健身房
13 母婴室
14 无障碍卫生间 / 后
   期可换成淋浴房
15 更衣室
16 办公区
17 开敞办公区

华贸中心·优客工场

项目背景

华贸中心·优客工场项目坐落在 CBD 核心商圈，建筑面积 6500 平方米。周围云集超 5A 智能写字楼、中央广场及商业等地标性建筑，拥有独特的国际资源优势。设计团队潜心探索设计的无限可能性，融合北京前卫大胆的创新精神，不断碰撞后产生出新的极富生命力的办公空间创意。

"设计是自然得出提炼，以一种纯几何方式出现的因素。"

——赖特

Project Background

Hua mao Ucommune is located in the core business district of CBD, with a construction area of 6500 square meters. It is surrounded by landmark buildings such as super 5A intelligent office buildings, central plaza and commercial buildings, with unique international resource advantages. The design team devoted themselves to exploring the infinite possibilities of design, integrating the avant-garde and bold innovation spirit of Beijing, and generating new creative office space with great vitality after continuous collision.

中国·北京 朝阳区西大望路甲6号

城市语境

华贸优客工场的位置正好在华贸商圈的边缘，基地前身是一处挑高近 10 米的巨大厂房，在城市更新的过程中，很多类似的城市旧建筑在城市的发展变革过程中亟待重新升级、定义其空间内容，以便承担适应城市发展的新使命。

我们通过加固和修缮旧建筑，在空间中搭建了两个夹层，分别承载不同的功能。联合办公是一种新型的办公形态，与以往的传统办公相比具有更大的自由度和选择性，"在生活中工作"是来形容自由创业者的工作和生活状态，因此在空间设计和内容设计上也需要直接反映这种新的空间特质。我们将首层定义为社交区域，二层和三层定义为工作区域。所有的生活部分，餐厅、咖啡吧、路演厅、美甲、健身以及对外出租的会议中心都在一层，相应的二、三层为可出租的办公区域。在内容的设计上，我们的着眼点不仅只放在空间内部，通常我们会考虑社区周边 200 米半径的商业，将它们视为联合办公所提供服务的一部分，我们的社交区域通常设计周边服务的补充或企业客户的刚需，更多地我们有意将空间规划为一种可以聚集周边流量和内部流量的场所，通过"服务升级"实现"空间升级"，更像是一种撮合商业。

联合办公是城市升级迭代的一支推动力量，建筑和空间的迭代更新从来都是从内容开始的变革。

Urban Context

Hua mao Ucommune is located at the edge of Huamao Business Circle. The former base is a huge factory building with a height of nearly 10 meters. In the process of urban renewal, many similar old urban buildings are in urgent need of upgrading and defining their spatial content in the process of urban development and reform, so as to undertake the new mission of adapting to urban development.

By strengthening and repairing the old building, we created two mezzanine levels in the space, each carrying different functions. Co-working is a new type of office, which has greater freedom and selectivity compared with the traditional office. "Working in life" is used to describe the working and living conditions of free entrepreneurs. Therefore, space design and content design also need to directly reflect this new space characteristics. We defined the first floor as the social area, and the second and third floors as the work area. All living areas, including the restaurant, coffee bar, roadshow hall, manicure, fitness and conference center for rent, are located on the first floor, and the corresponding office areas are available on the second and third floors. In terms of content design, our focus is not only on the interior of the space. We usually consider the businesses within a radius of 200 meters around the community as part of the services provided by co-working. Our social areas are usually designed to supplement the surrounding services or meet the needs of corporate clients. More importantly, we deliberately plan our space as a place where we can gather the surrounding flow and the internal flow, and achieve the "space upgrade" through "service upgrade", which is more like a matchmaking business.

Co-working is a driving force for urban upgrading and iteration. The iterative renewal of architecture and space is always a change from the content.

设计理念

整体空间采用极简设计手法，将具有独特空间体验和内涵的旧工厂改造成一个挑战传统办公理念的办公空间。6500平方米的办公空间由独立办公区、开敞办公区、前台接待、咖啡区、会议区等多个部分组成。本次设计以"新陈代谢"为核心理念，结合"孵化器"一词，将新兴的办公类型"孵化器"联想到另一种性质的孵化器——"巢"作为设计元素，并运用了现代设计手法进行元素提炼，以此为空间带来新的能量，将老厂房的"生命"延续，使其重焕生机。设计团队致力于将旧厂房打造成一个具有"场景性、互动性、开放性"特质的富有生命力的空间。

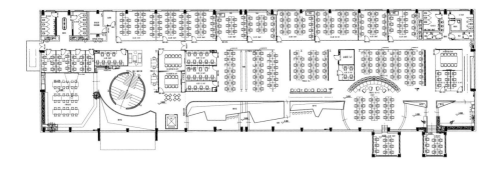

Design Concept

The overall space adopts minimalist design techniques, transforming the old factory with unique space experience and connotation into an office space that challenges the traditional office concept. The 6500 square meters office space is composed of independent office area, open office area, reception, coffee area, meeting area and other parts. The design concept with "metabolism" as the core, combined with the word "incubator", will be the new office type "incubator" associated with an incubator of another kind - "nest" as design elements, elements and using the modern design methods, to bring new energy for the space, will be the "life" continuity of old factory building, make its reinvigorated. The design team is committed to turning the old factory into a vibrant space with "scene-like, interactive and open" characteristics.

空间内容

围栏犹如鸟筑成的巢穴造型，一丝一丝地缠绕在一起，编织着富有无限创造力的未来。

空间中蜿蜒回环的楼梯，简洁有力的线条构造，充满东方韵律之美，以鸟巢的形式构建更具立体感的空间，达成动态的平衡，巧妙的楼梯设计象征着不受传统羁绊的生命活力。

Spatial Content

The fence is like a nest built by a bird, one by one intertwined, weaving a future full of infinite creativity.

Winding and circling stairs in the space, simple and powerful line structure, full of Oriental rhythm beauty, in the form of bird's nest to build a three-dimensional space, to achieve dynamic balance, clever stair design symbolizes the vitality of life unfettered by tradition.

在强调设计主旋律的同时，保留老建筑的结构。设计师将"巢"转化为极具张力的线条，通过简约的造型来形成空间、分布空间、组合空间、贯穿空间，灵活布局之中将老厂房转换成充满生命力的现代办公空间。

扭曲的线条形如"生命组织"DNA，串联着整个空间。凝练的设计语言创造出丰富的空间层次感和节奏感，生动多样的空间氛围与延展性并存。元素的运用与空间功能有机结合，使之在空间内处处体现，既不突兀，又充满艺术感，通过鸟巢抽象而成的回旋吊顶让空间有了延伸，交叠呼应，配合几何形瓷砖，获得多维的视觉效果。老厂房"生命"的延续，新空间"生命"的诞生，不同办公形态和产品内容的完美组合，不同材质的搭配，实现了一个具有独特魅力的办公空间。

While emphasizing the design theme, the structure of the old building is preserved. The designer transforms the "nest" into extremely tension lines, forms space, distribution space, combination space and penetration space through simple modeling, and transforms the old factory into a modern office space full of vitality in flexible layout.

Twisted strands of "living tissue" DNA connect the entire space. Concise design language creates a rich sense of spatial hierarchy and rhythm, vivid and diverse space atmosphere and ductility coexist. The use of elements is organically combined with the function of the space, so that it is reflected everywhere in the space, which is not obtrusion, but full of artistic sense. The circular ceiling made by the bird's nest abstract extends the space, overlaps and echoes, and coordinates with the geometric tiles to obtain multidimensional visual effects. The continuation of the "life" of the old factory building, the birth of the "life" of the new space, the perfect combination of different office forms and product content, the collocation of different materials, achieve a unique charm of the office space.

Project Background

Alibaba Innovation Center—Zhao wei Ucommune is located in Building C3, Zhaowei Industrial Park, No.14 Jiuxianqiao Road, Chaoyang District, Beijing. Here is Beijing's "second CBD", and "Beijing's new cultural landmark" 798 Art District adjacent. The former site is the state-owned Beijing Power Plant, electronic information industry base.

The core requirement of co-working, innovation and entrepreneurship space is highly adaptable, practical and functional space. Therefore, the design team arranged 300 independent offices and 500 open offices in the 5900 square meters space.

兆维 · 优客工场

项目背景

阿里巴巴创新中心——兆维 · 优客工场位于北京市朝阳区酒仙桥路 14 号兆维工业园 C3 号楼。这里是北京的"第二个 CBD",又与"北京文化新地标"798 艺术区相邻。该地前身是国营北京有线电总厂,电子信息产业基地。

联合办公、创新创业场地的核心需求是高适应性、实用性和功能性的空间。因此设计团队将这处面积为 5900 平方米的空间设置了 300 个独立办公工位,500 个开放办公工位。

中国 · 北京市 朝阳区酒仙桥14号

Urban Context

Many original buildings unable to adapt to the change and development of the city, and some old buildings and factories need new content to revitalize them. Zhao wei optimal workshop venue before the reform, have experienced electronic components factory, TV accessories factory, such as a variety of historical identity, along with the development of the city, around Zhao wei industrial park has been put into adjacent to wangjing business circle, lido hotel and the 798 Art District of large commercial area, optimal guest works as the office form of emerging, made up for this area is for entrepreneurs and freelancers, And the needs of customized office of small and medium-sized enterprises.

城市语境

许多原有的建筑已无法适应城市的变化和发展，一些旧建筑、旧厂房需要有新的内容使之焕发新生。兆维优客工场的场地在改造前，曾经历过电子元器件厂、电视机配件厂等种种历史身份，随着城市的发展，兆维工业园的周边已经形成毗邻望京商圈、丽都商圈和 798 艺术园区的大的商业区域，优客工场作为新兴的办公形态，弥补了这个区域对于创业者和自由职业者，以及中小企业定制化办公的需求。

运营者说

李晓牧_优客工场首席开发官

这些年东奔西走、海内海外，改造了 100 多个场地，从写字楼、商场、厂房到胡同小院、文物古迹，各色各样的房子汇聚了共享办公产品的一次次迭代更新，每一个产品从设计到施工、交付，就像培养孩子一样充满了期望与呵护，看着各个产品运营后给客户带来的舒适便捷，对于我们这些建设者来说也是最大的欣慰，希望我们未来能给城市的更新改造带来更多的期待和惊喜。

杨阳_优客工场成本总监

改造一个共享办公的环境，就是把有限的成本投入，变成无限的服务输出，形成有形的产品和可持续的服务链。重点在设计阶段控制成本水平，做到物尽其用，寸土成金。

场景营造

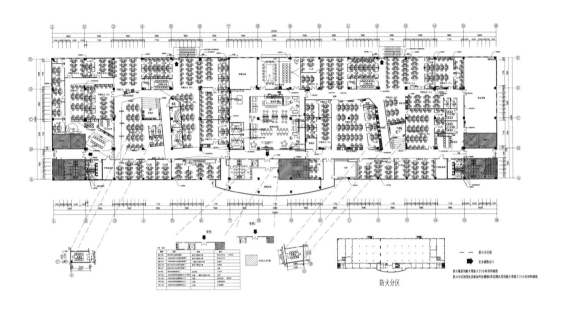

设计理念

整个场地的中央为共享休闲交流区，包括随处可站可卧的休闲区以及其他正式与非正式洽谈区和会议空间等。左右两侧主要是办公的形式，包括私人办公、独立办公间，会议室和开放办公区，通过合理布局满足多元功能需求又不会感觉拥挤。设计团队以"新陈代谢"为核心理念，以东南西北折纸游戏图案为抽象线条，辅助以六边形为基础元素，共同打造一个充满活力的图案游戏社区。温暖的大型木质 LOGO 仿佛海洋中的一艘航船。通过建筑构造、色彩、线条和观者的自我感知，打造一处集接待、洽谈、头脑风暴、共享咖啡吧、会议、路演于一身的全新多元功能体，实现"新陈代谢"核心理念下的需求与空间模式的动态平衡。

东南西北

1. 向心折

2. 四角向后折

3. 沿虚线向箭头方向折出折痕

4. 整理成双正方形对折，再撑开外层

5. 整理

6. 翻过来，将双手的拇指和食指插在四个尖角里，就可以玩了

Design Concept

The center of the whole site is a shared leisure and communication area, including the leisure area where you can stand or lie anywhere, as well as other formal and informal negotiation areas and meeting spaces. The left and right sides are mainly in the form of office, including private office, independent office, conference room and open office area, through reasonable layout to meet the multi-functional needs without feeling crowded. The design team takes "metabolism" as the core concept, takes the southeast, northwest origami game patterns as abstract lines, and assists the hexagon as the basic elements to jointly create a vibrant design game community. The warm large wooden LOGO resembles a ship in the ocean. Through the architectural structure, color, lines and the self-perception of the audience, a new multi-functional body integrating reception, negotiation, brainstorming, shared coffee bar, conference and road show is created to achieve the dynamic balance between the needs and spatial patterns under the core concept of "metabolism".

主体空间色彩主以黄蓝两色为主，黄色的轻快、活力与蓝色的冷静、理智，在空间中相互配合又相互"制衡"，营造着温暖而不失工作氛围的环境，形成一个充满生命力的有机空间。

以折纸图案为基底，分解出六边形元素，并相互结合，连续变形重组，并以花砖、地板、地毯、墙纸、玻璃贴等多种材质，虚实结合，在不同的区域以不同的形式出现，同时利用空间高度，跳脱出寻常的套路，放飞了创新思维的同时，也给空间带来了更多了可能性，拥有无限的活力。

在交通节点及主要流线上，精心设置的交流、独处、休息设施，也成为图案游戏的延续，为紧张的工作提供了各种方式的恢复空间。

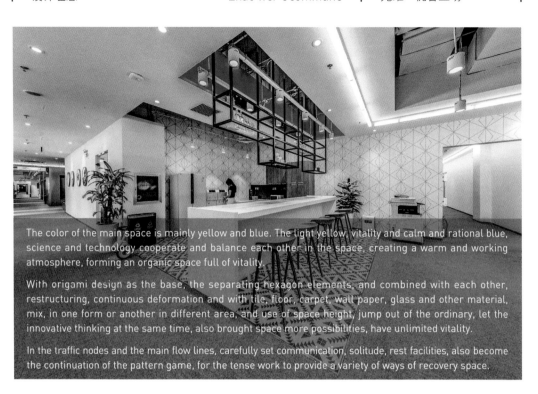

The color of the main space is mainly yellow and blue. The light yellow, vitality and calm and rational blue, science and technology cooperate and balance each other in the space, creating a warm and working atmosphere, forming an organic space full of vitality.

With origami design as the base, the separating hexagon elements, and combined with each other, restructuring, continuous deformation and with tile, floor, carpet, wall paper, glass and other material, mix, in one form or another in different area, and use of space height, jump out of the ordinary, let the innovative thinking at the same time, also brought space more possibilities, have unlimited vitality.

In the traffic nodes and the main flow lines, carefully set communication, solitude, rest facilities, also become the continuation of the pattern game, for the tense work to provide a variety of ways of recovery space.

空间形态

经过场景式设计的会议室，平面的空间变得立体化、生动化，可以让每一次的讨论或是思考充满兴致。

在交通节点及主要流线上，精心设置的交流、独处、休息设施，也成为图案游戏的延续，为紧张的工作提供了各种方式的恢复空间。

柔美的色调、通透的玻璃隔墙，室内洋溢着欢快与轻松的氛围。在保证开放视觉的同时也不忘均衡私密、实用的需求。独特的配色和随处可见的优客标志丰富了空间色彩，同时更是塑造出一个独具魅力的办公空间。

Spatial Form

After the scene design of the conference room, the plane space becomes three-dimensional, vivid, can let every discussion or thinking full of interest.

In the traffic nodes and the main flow lines, carefully set communication, solitude, rest facilities, also become the continuation of the pattern game, for the tense work to provide a variety of ways of recovery space.

Soft and beautiful tone, transparent glass partition wall, indoor permeated with cheerful and relaxed atmosphere. In order to ensure the open vision at the same time do not forget the balance of private, practical needs. Unique color matching and ubiquitous Youke logo enrich the space color is also to create a unique charm of the office space.

南京高铁·优客工场

项目背景

中国·南京 雨花台区六朝路18号

南京高铁·优客工场项目位于南京市高铁广场的2层，直接面对高铁站，交通位置非常便利。由于连通了南京的高铁，所以这里也可以被视作是南京的一个窗口，可以向来往的人们展示南京的历史文化以及人文气息。此次项目的方案在融合当地特色的同时符合现代文化环境，通过多种功能的联动，打造出集办公、休闲、娱乐、学习等多功能的联合办公空间。

Project Background

Nanjing SHSR Ucommune project is located on the 2nd floor of Nanjing high-speed railway plaza, directly facing the high-speed railway station, and the transportation location is very convenient. Because it is connected to Nanjing's high-speed rail, it can also be regarded as a window of Nanjing, where it can show the history, culture and cultural atmosphere of Nanjing to people who come and go. The plan of this project is in line with the modern cultural environment while integrating local characteristics. Through the linkage of multiple functions, it creates a multi-functional co-working office space that integrates office, leisure, entertainment, and learning together.

设计理念

南京高铁·优客工场的使用面积大概3000平方米，其中包含了87个独立办公室，共704个办公位；38个开放工位，3个10人会议室，2个8人会议室，以及两个半开放的可按需转换为路演厅的10人会议室；集合水吧与休息区、打印、休闲办公等空间。整个空间里，不仅能满足多种办公方式，还有配合办公使用的各类公共空间，打造多元化的办公环境。配色主要以樱花粉为主色调，黑白灰为辅助色，草绿色为点缀色，整体氛围轻松时尚，清新素雅。

Design Concept

Nanjing SHSR Ucommune covers an area of 3000 square meters, including 87 independent offices with a total of 704 office Spaces. 38 open stations, 3 meeting rooms for 10 people, 2 meeting rooms for 8 people, and two semi-open meeting rooms for 10 people that can be converted into roadshow halls on demand; Water bar and rest area, printing, leisure office space. In the whole space, not only can meet a variety of office methods, but also various public spaces with office use, to create a diversified office environment. The matching colors are mainly cherry blossom pink, black, white and grey as auxiliary colors, and grass green as ornament colors. The overall atmosphere is relaxed and fashionable, fresh and simple and elegant.

水吧与前台作为一个整体的方盒子来设计，顶地面的衔接增加了整体性。吧台侧面采用背漆玻璃材料，可用于张贴便签、通知、告示等，也可以涂写 DIY。水吧台也可作为小型会议室，临时办公或者员工随意涂写释放灵感的区域。休闲区的背景板选用木制的洞洞板，背景板上的吊柜位置可以随意调节，也可以粘贴、放置或者悬挂一些有趣的小物件，增加空间的灵活性。

The water bar is designed with the front desk as a whole square box, and the connection of the top floor increases the integrity. The side of the bar is made of lacquered glass, which can be used for posting notes, etc., and can also be painted with DIY. The water bar can also be used in small conference rooms, temporary offices or areas where employees scribble to release inspiration.The wooden hole board is selected for the background board of the leisure area, and the position of the hanging cabinet on the background board can be adjusted at will, and you can also paste, place or hang some interesting small objects to increase the flexibility of the space.

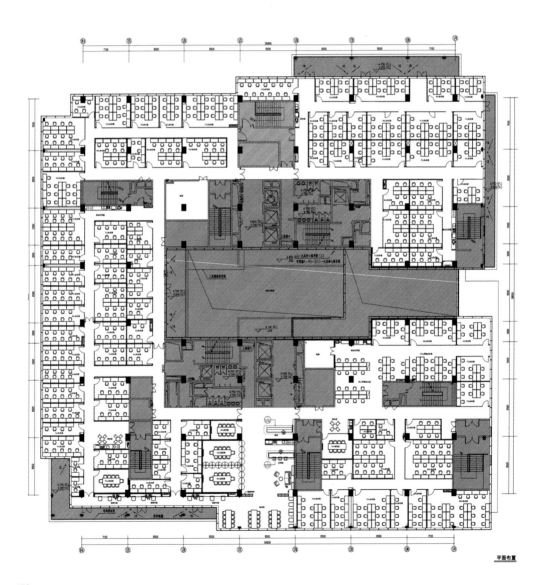

平面布置

空间场景

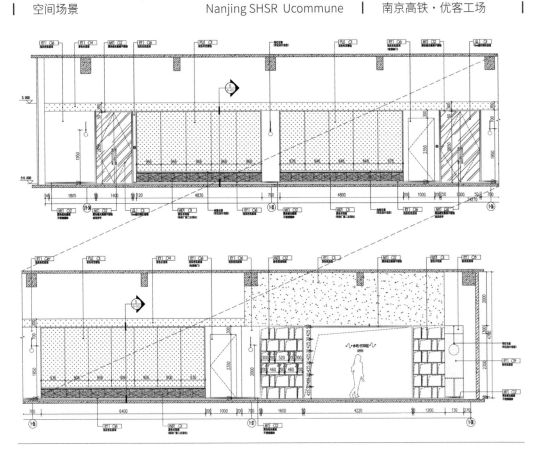

阳光 100 · 优客工场

项目背景

阳光 100・优客工场位于北京 CBD 的核心位置光华路上，地上地下部分建筑面积 12600 平方米，拥有 1432 个工位，是优客工场总部所在地，优客工场全国第一个社区，也是目前为止最大的社区，具备联合办公所有的办公、会议中心、大庆朗读书店、咖啡、健身、画廊、餐饮零售等商业全生态社区，也是城市旧空间升级迭代的第一个范本。

中国・北京 朝阳区光华路甲2号阳光100 D座

Project Background

Sunshine 100 Ucommune is located on Guanghua Road, the core of Beijing CBD, with a floor area of 12600 square meters and 1432 work stations. It is the headquarters of UrWork and the first UrWork community in China, as well as the largest one so far. It is also the first model of the urban old space upgrading and iteration. It is equipped with all the commercial ecological communities such as office, conference center, Daqing reading bookstore, coffee, fitness, gallery, catering and retail.

城市语境

互联网的发展推动着城市的变革，移动互联网的发展更是更新了人们的办公形式，在这种背景下社会资源再分配，最终需要城市空间的迭代更新以适应新的生产力的发展。人们不再被局限于固定的工位、办公室之中，取而代之的是在全球的网络中更加自由便利地移动，而联合办公就是这个巨大的网络中的节点，在这个节点中人们获得了更大的工作自由，更加便捷和舒适的工作环境，以及更广泛的社交和最大化的信息传递。工作和生活的边界变得模糊甚至相互交织，优客工场就是处于这种节点上的联合办公的杰出范本。

Urban Context

The development of the Internet promotes the transformation of cities, and the development of mobile Internet has even updated people's office forms. In this context, social resources are being distributed, which ultimately requires the iterative renewal of urban space to adapt to the development of new productivity. People are no longer limited to a fixed location, office, instead it is more freedom in the global network convenient movement, and joint office is the huge network of nodes, the nodes of the work people alive have greater freedom, more convenient and comfortable working environment, as well as the broader social and maximization of information transfer. Ucommune is an excellent example of co-working at a juncture where the boundary between work and life becomes blurred and even intertwined.

首层的大堂是北京唯一一处书店式大堂，也为地下空间的"大庆朗读"做设计语汇上的铺垫。

设计理念

阳光 100·优客工场的设计出发点，优先设计的是空间的内容，然后才是内容所引导的风格和设计语汇，以便在未来的运营中实现空间内容的良性循环，并且在内容的升级过程中可以逐步更新迭代。

阳光 100 优客工场总体分为两大部分，即办公区域和非办公区域。办公区域是包含 1～15 人不等的独立办公间和开放办公区域，以及为办公配套的会议室；非办公区域是包含为社区企业成员提供配套服务的各种商业空间，如书店、咖啡、餐厅、路演空间、供租赁的会议室、健身房、瑜伽房、直播间和录音棚等不断迭代内容的商业空间，让平台上的企业会员可以舒适地在"生活"中工作，以及在这种"生活方式"中寻找各种可"撮合"的商业机会。

Design Concept

The starting point of the design of Sunshine 100 Ucommune is to design the content of the space first, and then the style and design vocabulary guided by the content, so as to realize the virtuous cycle of the space content in the future operation, and can be gradually updated and iterated in the process of content upgrading.

The whole Sunshine 100 Ucommune is generally divided into two parts, namely office area and non-office area. The office area is an independent office with 1~15 people and an open office area, as well as a meeting room for the office. The office area is included for the community member companies to provide supporting services of all kinds of commercial space, such as bookstores and coffee, restaurant, road space, meeting room for rent, gym, yoga room, studio and studio iterated content such as commercial space, let the platform of enterprise members can be comfortable working in a "life", And looking for matchable business opportunities in this "lifestyle".

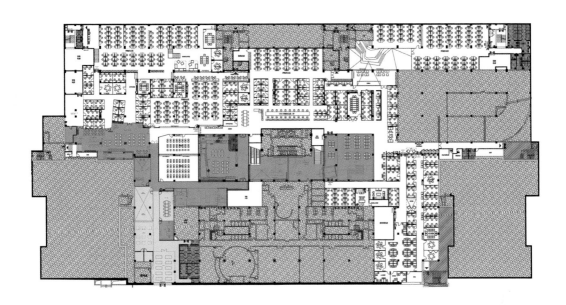

自助商店，不仅可以满足社区成员日常生活品的购买，同时也是销售企业产品的线下平台。

场景内容

用来进行直播的空间场景，以及场区内免费使用的工位。

共享厨房是给社区成员提供餐饮服务的场所，同时鼓励使用者进行寄存。

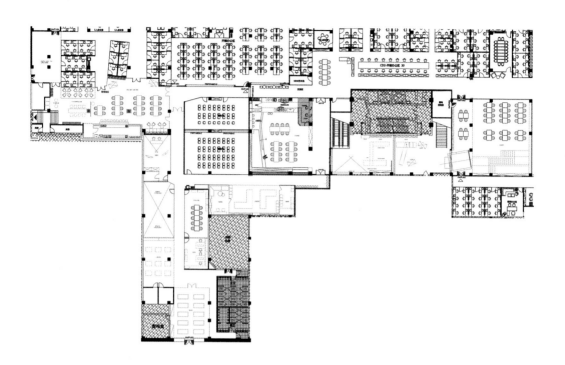

用木质台阶连接的两层空间，既是垂直
交通，又是特意设计的分享区域。

田涛_优客工场首席设计师

在优客工场经历的 6 年时光，不仅亲历了中国波澜壮阔的众创时代，也正在经历着城市更新的历史进程。通过对新的办公空间形式的探索和实践，为实现平凡人的创业作出点滴贡献，也切实地经历了一场创业的历程。

刘开颜_优客工场工程总监

优客工场工程管理紧密依靠战略合作单位的全力配合，进场前周密计划，过程中实行标准化安全质量进度管理，交付后及时回顾总结改进优化。不断自我完善和进化，凝聚成为强大的执行力，以年均交付 30 个场地的速度完成公司的全球布局和发展规划。

空间形态

大庆朗读	城市客厅
铁手咖啡制造总局	咖啡餐饮
本质咖啡	咖啡餐饮
两杯两杯咖啡	咖啡餐饮
胖妹面庄	网红餐厅
铁手咖啡前门店	咖啡餐饮
铁手咖啡杭州店	咖啡餐饮
果麦 2040 书店	品牌书店
布衣古书局	古籍书店
"塊"咖啡	咖啡餐饮
无瓦农场	休闲农场

新消费空间
New Consumption Space

Bookstore, Coffee Shop, Restaurant, Urban Living Room, Farm
Space and Scenario

The new consumption content focuses on vertical content, UGC communication, community operation and high appearance level space. Coffee, catering, bookstore and cultural space are all business forms with the characteristics of new consumption. More people start to grow grass online, pull weeds offline, and then spread the consumption and social mode. The space design also pursues the overall control and detailed description of the scene. With the support of good commercial products, it attracts the traffic and makes use of the Internet and scene marketing to create phenomenal commercial space products.

Operation

As a space and content operator, we provide our own space and cooperation methods for high-quality content, jointly incubate brand upgrading products, and create unique commercial products. At the same time, the main business and content in the community form a good mutual linkage, creating the uniqueness and humanization of community operation.

书店、咖啡、餐厅、城市客厅、农场

关于空间与场景

新消费空间注重内容垂直、UGC 传播、社群运营以及高颜值咖啡、餐饮、书店、文化空间，都是具备新消费特点的业态，更多的人开始线上种草，线下拔草，再传播消费和社交方式。空间设计也更加追求场景的整体把控与细节刻画，在良好的商业产品支撑下，吸引流量，利用互联网与场景营销，打造现象级商业空间产品。

关于运营

作为空间与内容运营商，我们提供自由的空间与合作方式，共同孵化品牌升级产品，打造独一无二的商业产品，同时社区中主营业务与内容相互形成良好的联动，创造属于社群运营的独特性和人性化。

大庆朗读

项目背景

"大庆朗读"位于北京市朝阳区光华路甲 2 号，是优客工场总部的一处网红社交场所。空间的前身是地下超市，后经改建为联合办公空间，使用面积 400 平方米。"大庆朗读"作为联合办公空间中的线下社交空间，为企业会员、周边写字楼及居民提供了一处集读书、咖啡、活动和零售为一体的空间场所。

中国·北京 朝阳区光华路甲2号优客工场

Project Background

"Daqing Space" is located at No.2 Guanghua Road, Chaoyang District, Beijing. It is a social place for Internet celebrities in the headquarters of Ucommune. The space used to be an underground supermarket, and was later transformed into a co-working space, covering an area of 400 square meters. As an offline social space in the joint office space, "Daqing Space" provides a space integrating reading, coffee, activities and retail for corporate members, surrounding office buildings and residents.

城市语境

城市的发展推动空间内容的迭代更新，一个地下空间从地库到超市，从超市到联合办公乃至一处新型的多元社交空间就是"大庆朗读"。在充斥着写字楼的CBD的核心位置创造一个充满温情的书店聚会场所，将周边的居民和工作者聚集在一起，把原本闲置低效的空间改造成新的空间形态。

Urban Context

The development of the city promotes the iteration and update of spatial content. An underground space from basement to supermarket, from supermarket to co-working and even a new multi-social space is "Daqing Space". Create a warm bookstore gathering place in the core of CBD filled with office buildings, gather the surrounding residents and workers together, transform the originally idle and inefficient space into a new space form.

设计理念

大庆朗读在功能上承载了优客工场联合办公的一部分，作为这种新型办公形式的接待空间、活动空间和社交空间而发挥作用。联合办公因互联网的发展而产生，新的社会资源再分配让每个工作者的工作环境更加的"多元"，同时城市土地的寸土寸金又促使我们在设计过程中充分考虑一个场地的效率，即按照不同的需求甚至时间节点来灵活组织空间功能。大庆朗读处于写字楼的地下部分，前身为超市，空间内容从超市升级为办公场所后首要解决的是如何吸引人从地面到地下来，因此这样的地下空间在设计上要足够的明亮，足够的生态和舒适，同时需要更多的空间运营内容。

Design Concept

Daqing Space is functionally a part of the joint office of Urwork, and plays a role as the reception space, activity space and social space of this new office space. Joint office due to the development of the Internet, a new redistribution of social resources let each worker's working environment is more "diversity", at the same time, the urban land sky-high prompted us in the design process fully consider the efficiency of a site, namely according to the different requirements and even time node to flexible organization space function. Daqing Space is located in the underground part of the office building, formerly known as the supermarket. After upgrading the space content from the supermarket to the office, the primary solution is how to attract people from the ground to the ground. Therefore, such underground space should be bright enough, ecological and comfortable enough, and at the same time, it needs more space to operate the content.

整个大庆朗读取代了从前联合办公的前台区域和等候区域，新的空间中除了这两项功能外还增加了"两杯两杯"咖啡、书店、潮玩以及画廊和网易云音乐线下分享等功能，在内容上进行了丰富。为了打破人们对地下空间昏暗、逼仄的印象，我们以一个书店的场景作为空间容器的背景，并且有意将书架设计出一种倾斜感，让人们感受到一个被书架"包裹"的空间。

通常情况下这里是一个可以享受咖啡和读书的地方，但每月几次的活动发布以及培训会是除了售卖书籍、咖啡外的主要收入来源，大庆朗读在前期设计就考虑了这些活动，因此一部分的书架是坐落在台阶之上的，这不仅是为了制造视觉上的层次变化，也是为了活动而设计的讲台，同时我们将视频设备和音箱嵌入进书架，在满足活动需要的同时，尽可能创造广告的收入。

墙面设计了绿植，发挥了空间挑高的优势，也强化了阳光透过玻璃顶照射进空间里的感受。"撸猫"是这里的网红概念，前台一侧的玻璃猫房为这里带来很多粉丝流量，人因爱自然而爱空间——空间的内容带来了流量，流量反过来也滋养了空间。

The whole Daqing Space has replaced the front desk area and waiting area of the former joint office. In addition to these two functions, the new space also adds the functions of Twoocuup Coffee, bookstore, tide play, gallery and NetEase Cloud Music offline sharing, which are enriched in content. In order to break people's impression of the dark and cramped underground space, we used the scene of a bookstore as the background of the space container, and deliberately designed the bookshelves with a sense of tilt, so that people can feel a space "wrapped" by the bookshelves.

Normally there is a place to enjoy coffee and reading a book, but a few times a month release of activities and training in addition to selling books, coffee is the main source of income, in the early stage of the design we have considered these activities, so part of the bookshelf is located in the steps above, this is not only to create visual hierarchy, It is also a platform designed for activities. Meanwhile, we embed video equipment and speakers into the bookshelf to meet the needs of activities and create as much advertising revenue as possible.

The wall is designed with green plants, which gives full play to the advantages of high space, and also strengthens the feeling of sunlight shining into the space through the glass roof. The glass cat room on the side of the front desk brings a lot of fan traffic here. People love space because they love nature — the content of space brings fans flow, and flow nourishes the space back.

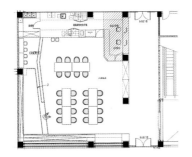

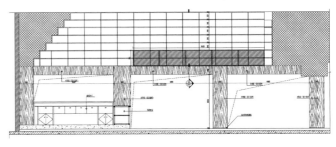

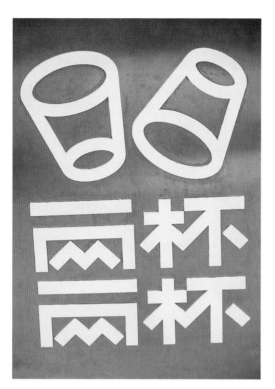

空间内容

与"旧物仓"的场景合作，链接了新旧两种时代的办公形式，时光在这里交错再现，年轻人在这里重温父辈的年代。这里变成了人们来大庆朗读打卡的地方。

在设计者的心中，"大庆朗读"在未来依然会随着内容的丰富生长、变化、不断地迭代更新。空间内容指导空间形式，这种设计思路会不断地持续下去。

Spatial Content

In collaboration with the "old warehouse" scene, the office forms of the old and new era are linked, where time is interwoven and the young people are reliving their parents' era. It has become a place where people come to read and punch in daqing space.

In the designer's mind, "Daqing Space" will continue to grow, change and iterate with the rich content in the future. Spatial content guides spatial form, and this design idea will continue.

主理人说

花小喵_大庆朗读常驻客人

喵、喵、喵

周孜柏_大庆朗读主理人

店是生活街区不可或缺的灵魂栖息地。若一个城市的书店日益增多，并且越来越有自己不可替代的特色，生活在其中的人也会对这个城市越来越眷恋。

我们就是旧铺改造而成，坐落在社区里，服务于社区。走进我们，你能在书店的某个角落发现很多"名片"。有附近的民宿，有好吃的在地小吃，有值得看的文艺演出或者画展。书店连接着这一片的人，加强了人与人的联系。书店是本地"信息聚集场所"。书店的意义不仅是卖书，更重要的是跟周围产生关联，努力成为社区所需要的一分子，让自身具有社会性。

我想书店是适合发生戏剧性故事的场景，生活虽不像电影总是充满戏剧性，但书店依然发生着种种动人的时刻，给人很多奇妙的感受。

铁手咖啡制造总局

项目介绍

空间张力：整体空间是由几个大小不一，建造年代不同的空间串连形成，各空间地面高低起伏变化有趣，有很强的空间塑造潜力。第一个空间是大厅区域，大厅是框架结构，层高 4.1 米，南侧有三个大的采光窗户，窗台较高，采光均好。大厅东侧有个狭长的通道进入到第二个下沉空间，这个空间是传统四合院的主屋，房屋是木结构，斜屋顶瓦房，之前被当作排练场的食堂，该屋正前方有一微小庭院空间，庭院小但采光充足，是整个空间仅有的户外空间。穿过"厨房"区域到"后厨"区，后厨空间层高 4.35 米，空间可塑性强。

物质的力、时间的力：大厅旧混凝土模板浇筑的印迹清晰可见，"厨房"区建筑是传统的梁木结构，年久失修亟待被清理、被发现，以重新散发古老魅力。被潮气及蚊虫侵蚀的木头柱、混凝土模板浇筑痕迹、木屋顶、旧瓷砖水泥贴面、青绿色旧窗户、旧水磨石地砖，我们认为所有场地的这些物质之间有很强的联系，彼此之间存在着特殊的力，它们在不同的年代被建造、修缮、连接、所用，经过时间的洗礼现在呈现出不同的状态及颜色，改造后与新空间巧妙结合，这一切是由时间作为连接而发生的变化和对话，新设计就是要强化这种连接，强化物质之间因新旧对比而产生的时间感，强调因空间变化而形成人对时间的感知。

Project Introduction

Spatial force: the overall space is connected by spaces of various sizes that were built in different years. The changing floor heights across spaces promise great potential for creativity. The first space is the lobby area, a frame structure with a ceiling height of 4.1 meters. The three huge skylight windows on the south, with their high windowsill, ensure ample lighting. A narrow corridor on the east leads to the second space, the sunken space. As the main house of the traditional Siheyuan, this space is a tile-roofed house with a wood structure, which was originally used as the cafeteria of the rehearsal spot. A small courtyard, the only outdoor area of the whole space, sits in front of the house and ushers in sufficient lighting. Through the kitchen area comes the back-kitchen area, whose ceiling height of 4.35 meters allows much room for creativity.

The forces of material and time: the marks of old concrete formwork pouring in lobby is clear. Kitchen area, a traditional wood beam structure, has not been properly maintained for years. It desperately needs to be cleaned so that its traditional charm can be rediscovered. The wood columns that have endured years of humidity and invasion of ants, the marks of concrete formwork pouring, the wood roof, the old ceramic tile on the concrete, the old turquoise window, the old terrazzo floor tile, all these elements, as far as we are concerned, are all interconnected, with a special kind of force flowing among each other. They were built, restored, connected and used in different times, all of which lead to how they look and what they are today. After renovation, they merge with the new environment, talking and interacting with each other through the bond stemmed from the time span. This new design seeks to intensify the connection, emphasizing on the passing of time by drawing comparison between different items as well as people's perception of time generated from the switching among spaces.

空间内容

如何将不同空间有机串连并形成稳定统一的空间而不显得凌乱；如何打造新型的、有趣的咖啡生活体验社区；技术上如何实现异形曲面混凝土吧台；如何有序引导共享际施工和本项目的交叉施工；如何实现 Metal Hands Coffee 工厂店"承上启下"的定位，如何提高 Metal Hands Coffee 工厂店的辨识度？

客人在不同的空间感受与切换、体验不同空间营造的"场"，人们在这个"场"里交流、碰撞产生更多的能量，这个"场"与共享际、共享剧场融合，共同在老旧胡同区域内形成力场，这个力场一定程度激活了胡同日常生活，也是我们对胡同文化生活方式的一次探索实践。

Spatial Content

How to organically integrate different spaces into a stable and unified one that does not appear to be messy. how to create a brand-new and interesting community for coffee lovers; technically, how to manufacture irregular-shaped and curved concrete bar counter; how to orderly carry out cross construction of 5LMeet and our project; how to deliver Metal Hands's intention for this factory shop to be a transitional link and how to make Metal Hands' factory shop a more recognizable destination.

By immersing in and switching from different spaces, customers are able to experience the vibes created by all these spaces, in which they use communications and exchanges to produce more energy. Interacting with 5L Meet and shared theater, all these vibes form a unique force field in old Hutong area, which, to some extent, stimulates the daily life and enriches cultural experience in Hutong.

项目信息

项目名称：Metal Hands Coffee 铁手咖啡制造总局
地点：北京市东城区南阳·共享际一层
委托方：Metal Hands Coffee 铁手咖啡
设计单位：力场（北京）建筑设计 Linkchance Architects
设计师：安兆学，候雪，于文婧，邵琳，张鹤
设计时间：2019 年 5 月～ 2020 年 1 月
竣工时间：2020 年
项目面积：300 平方米

主要材料：穿孔铝板、不锈钢板、竹钢、条纹玻璃、镜面板、水泥艺术漆

设计单位信息：力场建筑 Linkchance Architects
地址：北京市朝阳区酒仙桥万红路 10 号，恒通商务园 C 座无界空间 A3-3

本质咖啡

项目介绍

通过"框景"内的画面重新感受咖啡

本质咖啡坐落于北京市东城区鲜鱼口一座翻修历史建筑的西北角，店内面积约 30 平方米，是毗邻优客工场的一个重要商户。通过向街区居民、观光客提供优质精品咖啡，激活鲜鱼口周边商业区。

设计将焦点集中于光与阴翳的追逐游戏，在咖啡馆内塑造一系列的"框景"。屏蔽的咖啡调制区域被一层曲线的帷幕包围，并以同心放射布局的方式，为周边座椅形成优美的依托及入口廊道，引人入胜地传递着神秘与好奇感。

探索的旅程始于入口挑高的条形空间，一片悬于弧形桌面之上的轻盈冲孔铝板屏风围绕，展露出部分座位区。

巨大的临街窗户交汇处是咖啡购买的廊道。设计有意地将咖啡师安排在这个位置，以便为堂食与外带的客人提供服务。通过单点透视构图的方式塑造的窗口，形成了和咖啡师面对面交流的场景。在廊道末端的灯膜墙前，拿取手冲咖啡的画面被随之凸显。

空间里，OFFICE AIO 使用了柔和的品牌颜色、简约却又意蕴悠长的设计语言以及有趣且意想不到的空间旅程以打造独特的空间体验。这些元素引导客人关注种种精心安排的画面，调动着他们的感官（声觉、嗅觉、味觉）全方位地享受咖啡。

Project Introduction

Occupying a 30 square meter corner of a restored historic building, Basic Coffee is a key tenant of a co-working hub with a goal to activate the newly gentrified commercial zone of XianYuKou, Dongcheng District Beijing, through providing quality artisan coffee beverages to its neighbours and visiting tourists.

The design focuses on establishing a series of "framed views" into the cafe, by orchestrating a play of light and obscurity. This sense of mystery and curiosity is made possible by a concentric spatial organisation, originated from a concealed coffee making area, wrapped by a 'veiled' entry gallery and a curved seating area on the outermost layer.

The journey begins with entering into a thin slice of volume enveloped by a suspended aluminium perforated screen above the bar bench, revealing a partial view of the seating area.

Coffee order takes place at the junction between entry gallery and the large window facing the street, where the barista is strategically positioned to serve both dine-in and takeaway customers, sculpted as an exaggerated one-point perspective scene. A fissure is created adjacent to an illuminating wall, dramatising the display of pour-over coffee preparation.

To craft the passage, Office AIO designs with pastel brand colour, the minimal but profound design language as well as a playful and unexpected spatial journey. These elements yield an experience that allows the patrons to experience the space through observing a series of orchestrated movements, experience coffee-drinking through focusing their senses on sounds, olfaction and flavours.

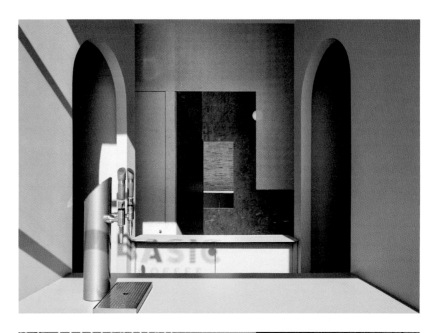

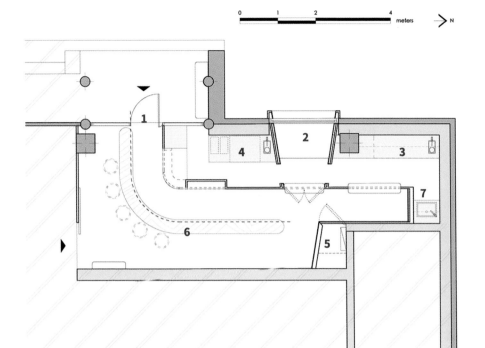

1 入口
2 外卖窗口
3 制作区
4 制作区
5 创藏室
6 座位区
7 消毒间

项目信息

项目名称：本质咖啡
地点：北京市东城区鲜鱼口
委托方：本质咖啡
设计单位：Office AIO
设计师：Tim Kwan
设计时间：2018 年
竣工时间：2019 年
项目面积：30 平方米
主要材料：钢板、石材、穿孔铝板

设计单位信息：Office AIO
公司网址：www.office-aio.com

两杯两杯咖啡

项目介绍

通道中的通道，拱顶下的拱

"两杯两杯"作为嵌入式咖啡运营服务品牌，希望在一家酒店主体空间交通节点处嵌入咖啡店铺，连接酒店大堂、步梯与其他商业空间，为主体空间内和临近胡同中的顾客提供精品咖啡服务。

我们在 10 平方米的狭小空间内，通过置入结构单元，满足精品咖啡制作功能的同时，延续品牌元素，吸引顾客停留并产生有趣的连接。旨在构建具有社区感、创造联结、多维互动的精品咖啡品牌。

店铺空间位于酒店一层临胡同一侧，其主要入口经由酒店大堂进入步梯间，与步梯对立，并将交通延伸至其他商业空间。临胡同一侧仅可留外卖窗口。

结合两杯两杯品牌基调和重叠的元素，依交通特性在空间中放置"重叠"的体块，通道中增加通道意象的结构，在拱顶下建立拱洞。

白色拱洞为钢板夹筋制作，让拱形看面更显轻薄。拱洞壳状结构由吧台上方搭向一侧墙体，墙面钢板翻折出高度适中的座面和吧台外侧的上旋翻板，为顾客争取了在"通道"内停留的时间，两杯咖啡的时间。

拱洞尽头，品牌色框的玻璃窗连接室外，一旁可开启的窗扇为胡同内经过的人提供外卖服务。

吧台整体由不锈钢板包裹，在顶部设置的反射光的作用下，模糊了吧台内外的界线。吧台顶部四个连续的拱面垂直向连接拱洞顶面，并停止在拱洞外表面，体块顶面亦与拱洞顶点相切，其内部框架也支撑加固了拱洞的壳状结构。连续拱面在另一侧与墙面上的四盏壁灯同心，让均匀受光的拱面与墙面有了更加明显的边界。

Project Introduction

As the embedded coffee service brand, "twoocuup" expects to embed the coffee shop at the conjunction of the main passages of a hotel, in order to connect the lobby, escalators and other business space, which is able to provides the premium coffee services for the customers in the main space and the people nearby.

The structure units are established in the narrow space, 10 square meters. However, it could not only satisfy the premium coffee—making process, but expand the brand elements, which is the link to attract new guests. Thus, it is helpful to make a community coffee brand with linking to creation and various interaction.

The coffee shop lies on the first floor at the side of Hutong. The main entrance is opposite the escalators through the lobby, and it also expands to other business areas. Also, the side window by Hutong can be the takeaway point.

Twoocuup is built with the elements of brand and overlap, and the overlap blocks are put in the room, which increases the passage looked structure with the arch under the arches.

The white arch is made of steel sandwich (keel), which looks thinner and lighter. The shell-liked arch is built from the top of bar counter to the other side of wall. Then the steel plates fold the appropriate height of seat, with the outside folding board of bar counter, which could extend the time for customers, time for two cups of coffee.

At the end of the arch, there is a glass window connects the outside, which could provide the takeaway service.

The bar counter is surrounded by the stainless steel plates, with the reflecting light from the top, the frontier looks blurry. the consecutive arches is vertical to the top of the arch, and it stops at the outside of the arch. Also, the top of block is tangent with the arch peak. The inside structure also supports the shelled structure. The other side of consecutive arches is concentric with the four wall lamps, so the even light distinguishes the frontier between the arches and the wall clearly.

Urban Context

The ends of Beijing alley is a magical place, Hutong is life, Hutong is outside the city, residents' sense of belonging is between step, when we put a small coffee shop in alley mouth, is to bring coffee to Hutong, brings residents, crept into the community of the home in the city of element, the fuzzy line between business and community, let the scene become warm and is novel.

城市语境

北京的胡同口是一个神奇的地方，胡同里就是生活，胡同外就是城市，居民的归属感就在一步之间，当我们在胡同口放置一个小小的咖啡厅时，就是希望把咖啡带给胡同，带给居民，城市的元素悄悄进入家里的社区，模糊商业与社区的界限，让场景变得温暖又新奇。

主理人说

两杯两杯咖啡北新桥店

"两杯两杯"致力于为多场景打造创意有趣的一站式饮品解决方案，同时提供精品、跨界、社区化的运营服务与内容。北新桥店是"两杯两杯"的首家精品咖啡门店，位于念念行旅一层，由"未来以北"操刀建筑设计，将咖啡、阅读、旅宿进行了深度结合。让喝咖啡这件小事更有趣，一起喝两杯。

未来以北建筑设计事务所

我们在 10 平方米的狭小空间内，通过置入结构单元，满足精品咖啡制作功能的同时，延续品牌元素，吸引顾客停留并产生有趣的互动。旨在构建具有社区感、创造联结、多维互动的精品咖啡品牌。

很难想象一个没有座位的咖啡空间如何让客人悠闲地停留，我们设计许多构件和装置，让不可能成为可能。

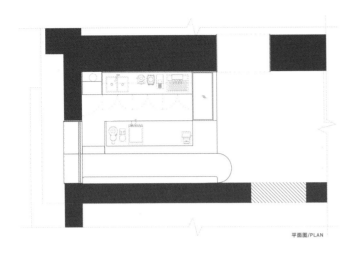

平面图/PLAN

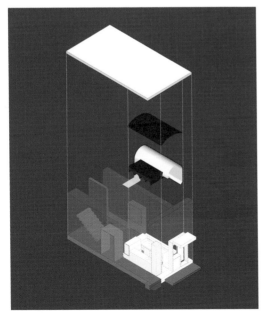

项目信息

项目名称：两杯两杯咖啡空间设计

地点：北京市东城区念念行旅酒店一层

委托方：两杯两杯咖啡

设计单位：未来以北工作室

设计师：金波安，李泓臻，罗霜华

设计时间：2019 年 8 月～ 2019 年 9 月

竣工时间：2019 年 11 月

项目面积：10 平方米

主要材料：不锈钢、松木多层板、钢板

设计单位信息：未来以北工作室

公司网址：www.fon-studio.com

胖妹面庄

项目介绍

启发自西南中国的地形地貌

OFFICE AIO 设计将店铺全新地接地气改造，把传统的面馆转变为只能在其中才能享用到正宗的重庆小面的美味以及氛围。OFFICE AIO 从摄影师 Tim Franco，Nadav Kander 和 Mark Horn 的作品中汲取了灵感，这些作品记录了快速城市化过程中多层次垂直的城市中的生活。

面馆分为两部分，后面的区域作为厨房，其余部分作为用餐区。一块下半部分朦胧，上半部分通透的玻璃，穿过两个区域之间，为繁忙的餐厅创建了系统的服务流程。尽管现有店铺是原来面积的两倍，但厨房的空间却比以前增加了三倍，设计师力求在用餐座位数保持不变的情况下，为厨房工作人员和顾客提供充足的空间和舒适的体验。借助悬挂的可互换红色菜单，个性十足，引人入胜，吸引了路边来来往往的人群。

高高低低起伏的地台采用灰色水磨石的材质，一直延续到餐厅同样是现场浇铸的水磨石。灰色水磨石以细腻的肌理纹路展现出非凡的格调，干净纯粹，置身于此，仿佛周遭的一切变得安静，水磨石的粗糙质感，让空间多了一份沉稳，从而达到一种高级的平衡感。不同长度的桌子散落在地上，并配有折叠凳子，使顾客产生在重庆特有的 3D 魔幻城市景观中熟悉的街头用餐的感觉，充满了人间烟火的味道。

Project Introduction

OFFICE AIO design transforms the store into a new, down-to-earth transformation, transforming the traditional noodle shop into a place where you can only enjoy the authentic Chongqing noodle taste and atmosphere. OFFICE AIO drew inspiration from the work of photographers Tim Franco, Nadav Kander and Mark Horn, which document life in the multi-layered, vertical city amid rapid urbanization.

The noodle shop is divided into two parts, with the rear area serving as the kitchen and the rest serving as the dining area. A glass that is opaque in the lower half and transparent in the upper half, passes between the two areas and creates a systematic service flow for the busy restaurant. Despite doubling the size of the existing store, the kitchen has tripled in size, aiming to provide ample space and a comfortable experience for both kitchen staff and customers while maintaining the same seating number. With its interchangeable red menu hanging, it is full of character and charm, attracting the people who come and go along the road.

The undulating platform is made of grey terrazzo, which is also cast on site. Gray terrazzo with fine texture lines to show a special style, clean and pure, placed in this, as if everything around become quiet, the rough texture, of terrazzo makes the space more calm, so as to achieve a senior sense of balance. Tables of different lengths are scattered on the floor, complete with folding stools, giving customers the familiar feeling of street dining in Chongqing's unique 3D magical urban landscape, filled with the taste of human fireworks.

苑杰_胖妹面庄主理人&摇滚音乐人

经营胖妹面庄之外，实际上我最本质的热爱是音乐，多年前曾组建了自己的重金属摇滚乐队，现在依然在为热爱的事业努力着。在设计中，我希望这是一个有情感的场景，高起的地台，暗示着重庆的地貌与文化，也希望这里像一座舞台，有一天这里也能成为我们的舞台，实现自己的梦想。面馆，舞台，山城，在一个小小的空间里，让设计师完美地呈现。

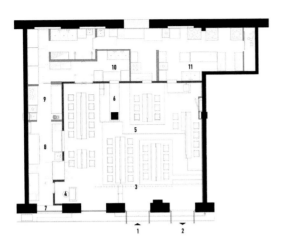

1　入口
2　出口
3　菜单板
4　点单台
5　座位 / 地台
6　出餐台
7　外卖窗口
8　饮品吧
9　甜品厨房
10　冷餐厨房
11　热餐厨房

项目信息

项目名称：胖妹面庄
地点：北京市东城区念念行旅酒店一层
委托方：胖妹面庄
设计单位：Office AIO
设计师：Tim Kwan，Isabelle Sun
设计时间：2018 年 12 月～ 2019 年 3 月
竣工时间：2019 年 11 月
项目面积：80 平方米
主要材料：不锈钢、大古水泥

设计单位信息：Office AIO
公司网址：www.office-aio.com

铁手咖啡前门店

项目背景

铁手咖啡前门店位于北京市东城区西打磨厂街口共享际内，是一幢极具风格的民国时期建筑，青砖灰瓦，门前的绿化景观为建筑平添了几分生机，西打磨厂街本就是一条兴盛于清末的商业街区，曾经大量商贾、会馆、饭庄、旅店云集于此，近百年的沉寂后，经过统一修缮与改造，整条街区焕发新的生机，大量新消费品牌正在这片充满底蕴与活力的区域不断生长，创造出一个属于新时代的前门商圈，让历史建筑重新焕发光彩。

中国·北京 东城区西打磨厂街228号

Project Background

Metal Hands Coffee Qianmen is located 5Lmeet in west polishing factory blocks, Dongcheng District, Beijing, is the very style of the Period of the Republic of China building, the green landscape in front of it added a few vitalities for the building. The west polishing factory street which is a thriving commercial district in last Qing Dynasty, has a large number of merchants, hall, restaurant, hotel gathers hereat. In one hundred years silence, after unification of restoration and reconstruction, the whole block is full of new vitality, a large number of new consumer brands are constantly growing in this area full of heritage and vitality, creating a new area of Qianmen business district, historic buildings have been revitalized again.

历史的透视

改造前的建筑是一栋传统的青砖建筑，内部的木质楼梯以及楼板已经因年久失修而失去了使用功能。改造时，我们希望尽可能地保留建筑原本的材质与肌理，让新的新，旧的旧，从而产生一种新旧对比，这种具有冲击性的对比能够让空间激起使用者与空间历史的对话。外立面尽量保留了建筑原本的面貌，通过一个开敞的庭院进入咖啡厅，就开始了一场与旧时光的相遇。

进入咖啡厅，从吧台向空间深处望去，就会看到三个层层相扣的拱券结构，第一层是不锈钢板拱券，在结构上支撑起原本老旧且存在安全隐患的墙面；第二层是混凝土拱券，我们在原始建筑的基础上涂刷了水泥漆，保持老房子岁月的质感；第三层是保留的外立面一致的青砖拱券。三层拱券的透视结构，让人仿佛沉浸在一种被历史包裹的氛围之下，一切旧的东西都以一种新的姿态呈现。

透过层层拱券，这种几何的透视感被加强，在视线上空间会聚焦成为一个类似于时空隧道的场所，那一抹打在咖啡杯上的阳光，已经分不清是现在的阳光还是过去的阳光。

The Perspective of History

The building before the renovation was a traditional brick building, and the internal wooden staircase had lost its function due to the disrepair of the floor. During the renovation, we hope to preserve the original material and texture of the building as much as possible, so that the new is new and the old is old, to produce a contrast between the new and the old. This impactful contrast can make space arouse the dialogue between users and the history of space. The façade retains the original appearance of the building as much as possible. Entering the cafe through an open courtyard, it begins an encounter with the old days.

When you enter the coffee shop and look into the depth of the space from the bar, you will see three interlocking arch structures. The first arch is a stainless steel arch, which structurally supports the old wall with potential safety hazards. The second arch is a concrete arch, which is painted with cement paint based on the original building to maintain the texture of the old building. The third one is a brick arch as same as the original facade. The perspective structure of the three arch makes people seem to be immersed in an atmosphere wrapped in history, and all the old things are presented in a new way.

Through the three layers of the arch, this geometric sense of perspective is strengthened, and space will focus on a place similar to a time and space tunnel, and the sunlight on the coffee cup can no longer tell whether it is the sunshine of the present or the sunshine of the past.

粗犷的对比

通往二层的楼梯间区域，不仅仅是一个交通空间，用一种既围合又通透的形式，将人们引导进入二层更加开阔的区域。设计采用了简洁的手法、极具工业感的混凝土墙面与现代感的金属板，肌理与光滑、粗犷与精致形成强烈对比，在这个非常小的空间内，材质对比的冲击效果被充分发挥，金色的"漂浮"楼梯和原色的水泥质感形成了强烈的冲击力。对比不仅能突出新与旧之间的关系，更是能让两种材料之间产生对话，从而让人与空间进行对话，产生思想与共鸣。

Rough Contrast

The staircase area leading to the second floor is not only a traffic space but also guides people into a more open area on the second floor in a form that is both enclosed and transparent. The design uses simple techniques, industrial concrete walls, and modern metal panels. The texture contrasts sharply with smoothness, roughness, and refinement. In this very small space, the impact of material contrast is fully utilized. The golden "floating" staircase and the cement texture of the original color form a strong impact. Comparison can not only highlight the relationship between the new and the old but also make a dialogue between the two materials so that people can have a dialogue with space.

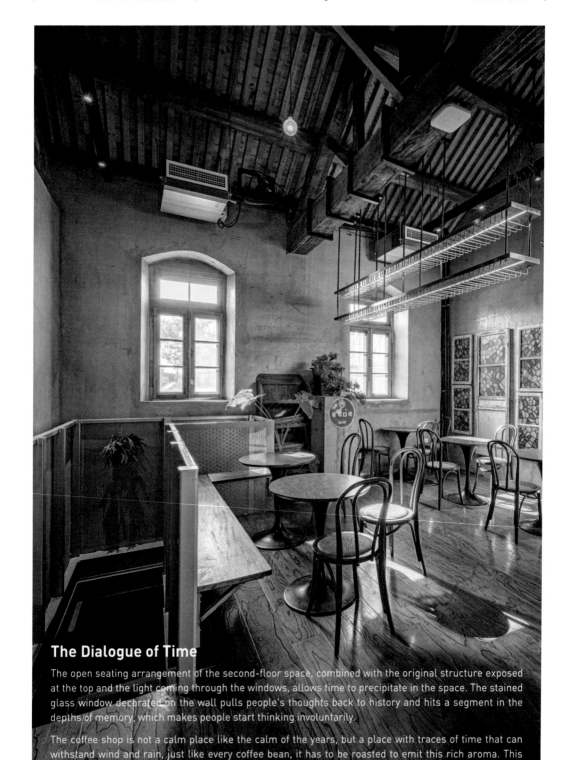

The Dialogue of Time

The open seating arrangement of the second-floor space, combined with the original structure exposed at the top and the light coming through the windows, allows time to precipitate in the space. The stained glass window decorated on the wall pulls people's thoughts back to history and hits a segment in the depths of memory, which makes people start thinking involuntarily.

The coffee shop is not a calm place like the calm of the years, but a place with traces of time that can withstand wind and rain, just like every coffee bean, it has to be roasted to emit this rich aroma. This kind of transformation, which makes a strong contrast and conflict between the new and the old, is also like coffee, which can stand the test of time.

时光的对话

二层空间的座位开敞布置，结合顶部裸露的原始结构，以及窗户透进来的光线，让时光在空间中沉淀。墙面装饰的彩色玻璃花窗将人们的思绪拉回到历史的某个节点，直击记忆深处的某个片段，让人不由自主地开始思考。

咖啡厅不是岁月静好般的风平浪静，而是能经得起风雨打磨的一种有时光痕迹的场所，就像每一粒咖啡豆，都要经历烘焙，才能散发出浓郁的香气。这种让新与旧强烈对比、冲突交织的改造手法，也像咖啡一般，能够经得起岁月的考验。

主理人说

丁老师_铁手咖啡主理人

铁手作为一个扎根于胡同的北京本土品牌，在打造门店空间的选择上与共享际坚持的商业空间运营理念不谋而合。合作契机源自我们的第四家门店——前门店，也是铁手第一次走出深藏的胡同，尝试"小而美"之外的运营模式，呈现了我们心中的"北平小品"。而这个设想在落地时获得了共享际从理念到空间再至运营的全面支持，并在碰撞中激发了更多灵感，对空间进行了完善。共生、成长、突破、完善是与共享际多个物业合作后的最大感受与收获。

铁手咖啡杭州店

穿越与想象

一位南宋侠客，在大雨中隐憩在宋代街坊中的一间茶肆，消失在江南小巷中。从项目最初，类似的场景无数次浮现在设计团队的脑海中，也贯穿于我们与业主的数次交流中。在无限变迁的时空与场景中，历史与人的短暂交会浓缩在铁手咖啡饮食文化空间中，进行了一场穿越古今的浪漫想象。

Time Traveling and Fantasy

Assuming we are experience time traveling to the ancient Song Dynasty, imagine a ranger hiding in a tea shop under the heavy rain. He was disappeared into the hazy night. From the beginning of the project, similar fragments of fantasy appeared several times in our minds. Meanwhile, the similar scenario of ancient Song's life scenes have also appeared several times in the conversation with our clients. We wants to create a brief encounter between the ancient and present, through the catering culture.

Project Background

This project is located in Dajing Alley, Hangzhou City, which connects the densely crowded Zhongshan Middle Road and Hefang Street. This alley has been located in the bustling core area since it was called Wushanfang back in the Southern Song Dynasty. The store is the pilot Metal Hands coffee shop runs in Hangzhou City, which is a risky decision to the client. How to integrate this domestic specialty coffee brand from northern China into this historical City in south? It was the core challenge for the design team and the client either. Considering the preservation and integration with the historical communities. We decided to abandon the sense of modern and industry, restore the elegant and leisurely life scene of the Song Dynasty. To design a wabi-sabi boutique coffee shop in this modern and bustling commercial spot.

项目背景

项目位于杭州市大井巷，连接人流密集的中山中路和河坊街，这条百米小巷从南宋时期名为吴山坊的时候就已然地处繁华核心地带，也是铁手咖啡在杭州的第一家门店的选址。如何将致力于本土精品咖啡的精品品牌店融入充满历史人文气息的江南小巷是设计团队的核心挑战。我们决定放弃现代和工业风格，而是去还原宋代淡雅闲适的生活场景。在现代的繁华商业场所中设计一个侘寂风格的精品咖啡馆。

中国 · 杭州　上城区大井巷81号

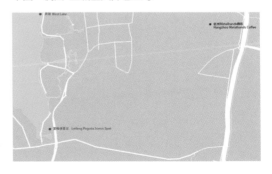

改造与重塑

宋人绘画作品中有非常多对于生活场景的描绘，例如
"春宴图""秋堂客话图""秉烛夜游图"等。围绕
着宋人格物致知的生活理念，在设计规划中，我们
尽可能地将不同功能的空间打造成沿街开放的多个
宴客场景，一物一景都经过精心的考量和设计，方
便空间内外的人进行细致入微的观察。

首层将原立面后退 1.2 米，以便在此进行景观造景。
外立面更换为可开放玻璃门窗，吧台台面向外突出，
形成内外空间的互动关系。

Renovation and Remodeling

There are many depictions of daily scenes in
the artworks of Song people, such as "Spring
Banquet" "Guestfeasting in Autumn" "Night
Tour" and so on. Regarding the delicate lifestyle
and philosophy of Song's people, we delicately
designed each program to mimic multiple banquet
scenes in Song's artwork , and open it along the
street for the viewer to look it up closely. The
overall details in the interior have been carefully
designed to facilitate the aesthetics of the space.

The original façade is set back by 1.2 meters for
display purpose. The north façade of first floor is
replaced and glazed with openable glass doors
and windows. The bar counter extends to the
streets, creating an potential interaction between
internal and external.

首层西侧三跨立面改为折叠门，成为店铺
的主要入口。西侧折叠门可全部向小院打
开，最南侧则改为落地玻璃。二层开放空
间的实体墙面与老窗户全部改为落地窗，
未来二层举办活动时会更加明亮开放。

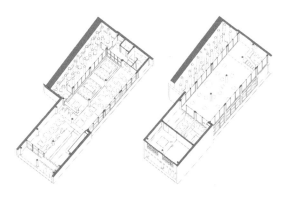

The three-span facade on the west side
becomes the main entrance of the store,
it were replaced with the folding door. The
folding doors are able to open up to the small
courtyard. The solid walls and windows of
second floor were also replaced with floor-to-
ceiling windows. In which, it will bring more
lights when events are held on in the future.

铁手宴客图

作为铁手在杭州的第一家实验型店铺，我们与业主交流后希望整体空间能更好地结合当地城市特点，整体呈现一个沉静且富有禅意的空间，通过空间视觉与空间氛围两个维度凸显人文与历史气息。

吧台由 100 毫米 ×100 毫米的木方错层穿插搭建而成。台面一侧的客人可以近距离观察咖啡师的工作行动。

铁手咖啡首层主要以接待区域和操作区域为主要功能。除了满足消费者与工作人员空间流线的合理性，在排布空间时我们没有选择常规客座利用率最高的方式，而是反其道而行之，尽量减少座位，不使用过于现代且棱角分明的桌椅家具。参考宋代画作场景当中常出现的软塌与坐墩，闲散舒适的姿态跃然纸上。在人们品鉴咖啡时，也拥有更通透开阔的视野去观察室内外景观。

The Feast of Metal Hands Coffee

Because it is the pilot store in Hangzhou, we share the same idea that the store should completely integrate the characteristics of the local specialty. Our concepts is to design a space with sense of calmness and Zen. Emphasizing the humanistic and historical atmosphere through the vision and it's atmosphere.

The bar counter is constructed by wood cut-stock, this share counter promote the customers and baristas interact with each other.

The first floor are divide into reception area and operation area .Besides to satisfy the efficient circulation between the consumers and staff, we did not choose the method with the highest utilization to arrange conventional seats and table in reception area. Instead, we minimized the quantity of the seats and tables. Inspired by the Chinese traditional furnitures appears in Song's paintings, the relaxed and comfortable posture of the Song people is vividly appears. We decided to arrange some lower table and seats for people to relax.

建筑东侧空置的天井处，我们使用阵列形式的树木作为景观装置，制造出一种神秘诗意的空间体验。从入口的小院处观察，咖啡店铺仿佛包围在自然环境之中。

At the vacant patio on the east side of the building, we array the trees as landscape installations to create a mysterious and poetic spatial experience. The coffee shop seems to be surrounded by a forest from the street side.

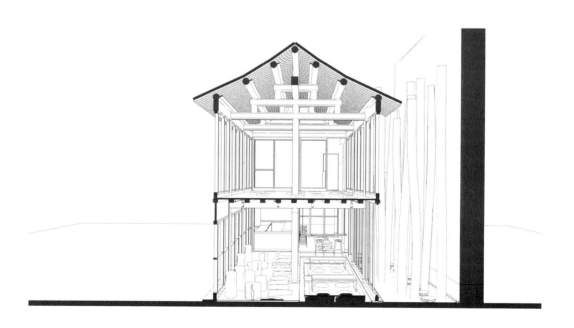

从一层空间南侧的回形台阶上到二楼空间，是完全开放的禅室冥想空间，兼具休息与活动功能。地面采用黑色不锈钢材质，将一侧大井巷古色古香的街道场景与另一侧中庭中沉静肃穆的景观装置同时反射进二层空间，使得禅室与周围的自然环境融为一体。时移世异，古人对于饮食生活的精致态度与文化精神，在铁手咖啡制造局的饮食空间内得以传承延续。

Going up to the second floor of the store, it is a completely open meditation space and Zen room. The ground is made of black stainless steel, reflecting the surrounding of antique architecture of Dajing Alley and the solemn trees installation on the other side. As time has change, the epistemological philosophy and culture are inherited and continued in the Metal Hands Coffee space.

主理人说

丁老师_铁手咖啡主理人

作为铁手在杭州的第一家实验型店铺，我们希望整体空间能更好地结合当地城市特点，整体呈现一个沉静且富有禅意的空间，通过空间视觉与空间氛围两个维度凸显人文与历史气息。尽量减少座位，不使用过于现代且棱角分明的桌椅家具。参考宋代画作场景当中常出现的软塌与坐墩，闲散舒适的姿态跃然纸上。

空间场景

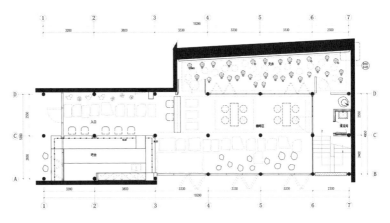

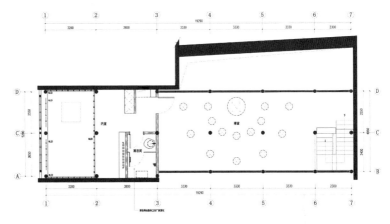

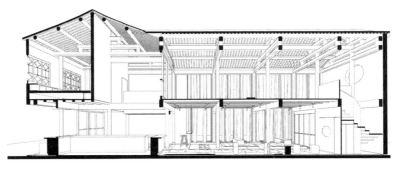

果麦 2040 书店

项目背景

果麦 2040 书店位于北京市东城区东外 56 号文创园内，书店位于主体楼三层的露台上，希望在保持原有咖啡店空间不变的情况下做共同经营。主理人希望能将果麦文化的所有出版物全面展示在书店内，使书店在咖啡店可使用空间得到充分激活与利用，又能变化不同的场景应对多种活动需求。书店与原有咖啡空间的关系是存于其中、独于其中的，旧空间里的新书屋。

中国・北京　东城区东外56号文创园

Project Background

The project is located in 56 Dongwai Industrial Park, Dongcheng District, Beijing. The bookstore is located on the terrace of the third floor in the main building. The bookstore hopes to do joint operation while keeping the original coffee shop. The manager hopes to fully display all the publications of Fruitwheat culture in the bookstore, so that the usable space of the bookstore in the coffee can be fully activated and utilized, and different scenes can be changed to meet the needs of various activities. The relationship between the bookstore and the original coffee space is existing in it, unique in it, the new bookstore in the old space.

设计理念

书店与原有咖啡空间的关系是存于其中、独于其
中的。原有空间整体为咖啡经营场所，中间是主
要营业区，书店希望对东西两头的空间进行使用
和改造，但是整体上不影响咖啡原有动线及客座
数。书店里咖啡馆场景重新规划平面布局后，西
侧为果麦文化最新出版书籍展示，中间廊道及东
侧为过往读物展示。既保留了原本经营良好的咖
啡馆，又丰富了原本单一的空间。划分方式不影
响原空间的经营动线，又将书店的属性独立以及
延伸。利用原建筑上的雨棚，整面的落地窗，设
计手法使用简单的框架装饰结构线条及室内灯光，
形成一整面大场景外立面的延展性及视觉面。

Design Concept

The relationship between the bookstore and the original coffee space exists in each other and also Independent of it. Original space overall is for coffee, in the middle of it is a major service areas. Bookstore wants to use the space that two head space and renovation, but it does not affect the original coffee moving line and number of seat, and the bookstore cafe scene, after planning layout, there displays the latest Guomai culture books on the west side,and the past reading in the middle corridor and east. The original well-run coffee space is retained in the middle, the contents enrich the original single coffee space. The way of division does not affect the operation line of the original space, but also makes the property of the bookstore independent and extended. Using the canopy of the original building, the whole floor to ceiling windows, and the simple frame to decorate the structural lines and indoor lights, form the ductility and visual surface of the facade of the whole scene.

设计者说

吴雪_共享际首席室内设计师

果麦 2040 书店，是在原有的咖啡店里生长出来的多功能展厅书店，通过灵活多变的活动书架装置，完成客厅式书店和多功能厅的自由切换，既是展厅又是多功能厅，用简洁灵活的设计装置，全面地展示了果麦文化传媒股份有限公司的 2000 多本出版物。

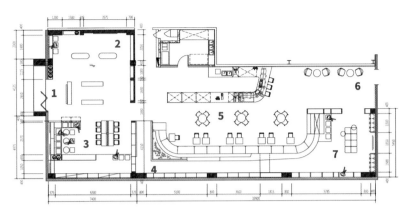

1 入口
2 书架陈列
3 阅读及活动区
4 书架
5 咖啡区
6 次入口
7 儿童阅读区

项目信息

项目名称：果麦 2040 书店
地点：北京市东城区东外 56 号文创园
委托方：果麦文化传媒
设计单位：优享与行策划咨询有限公司
设计师：吴雪
设计时间：2020 年 10 月～ 11 月
竣工时间：2021 年 2 月
项目面积：220 平方米

布衣古书局

项目背景

南阳·共享际是由 20 世纪六七十年代建成的中国演出集团公司的汇报演出剧场翻新改建而成的剧场项目。建筑分为两部分，一座近千平方米的剧场和一个文化社区，内部配套有商业、办公、工坊、公寓等空间。书店位于建筑内部剧场南侧走廊旁边的 34平方米的窄条空间，形状比较窄长，展示面在人流角度的走廊位置，共享际的设计师根据该位置的优势，在内部创造出与外围互动且有层次的书店空间。

中国·北京　东城区南阳胡同6号

Project Background

Nanyang 5Lmeet is a theater project renovated from the report performance theater of China Performance Group Corporation, which was built in the 1960s and 1970s. The building is divided into two parts, a nearly 1000 square meters theater and a cultural community which internal has supporting commercial, office, workshop, apartment and other space. The bookstore locates in a 34-square-meter narrow space next to the corridor on the south side of the theater inside the building. The shape is relatively narrow and long. The display surface is located in the corridor from the perspective of people flow. The designer of 5Lmeet created the book space interact with periphery according to the advantage of the location.

主理人说

胡同_布衣古书局主理人

认识共享际一年半，体会到一种在家的感觉。我的小店从设计到施工都交给了他们，属于典型的"拎包入住"。在这里，我们不仅仅是他们的租户，更是他们的朋友。从老总到员工，总是笑脸相对，一起来面对和解决遇到的所有问题。共享共生，他们的资源都无私开放给我们，共享际给我们提供的不仅仅是一块空间，更是一片天地。

挑　战

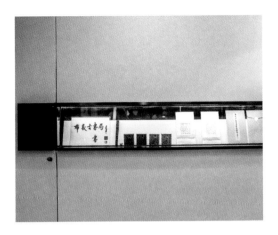

首先是在空间条件性的限制：书店是一个长 12 米、宽 2.8 米的狭长空间。要如何有效地在这个狭长的空间里打造一个具备展示古书、讲解古书，满足小空间使用的多样化及最大化利用率成了一个最大的难题。

从空间美学的角度来说：由于南阳·共享际已经开业运营，如何能与南阳·共享际现有的设计相融合，同时能在彰显自身"古籍"气质的基础上打造一个不拘于传统风格的古籍书店是关键，另外，如何调整原有外立面的通透性，既保证一定的展示功能，又避免书店空间一览无余非常重要。

Challenge

The first is the restriction of space conditions: the bookstore space is a narrow space of 12 meters long and 2.8 meters wide. How to create an effective display and explanation space for ancient books in this long and narrow space to meet the diversified use of small space and maximize utilization has become a biggest problem.

From the perspective of space aesthetics: since Nanyang 5Lmeet has been opened, how can it integrate with the existing design of Nanyang 5Lmeet and create an ancient book bookstore which is not in the traditional style on the basis of highlighting its own "ancient book" temperament is key How to adjust the permeability of the original facade, not only to ensure a certain display function, but also to avoid the bookstore space in a glance. that is important.

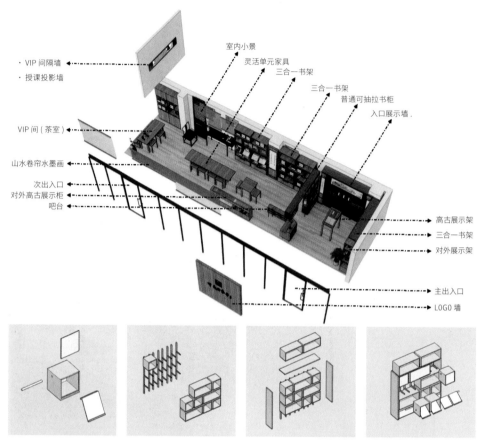

设计理念

设计师借鉴中式古典园林对空间划分的方式，移步换景，以小见大，巧妙地化解了狭长单调的空间带给人直白呆板的不适感受，将空间划分为三个虚实有异、分分合合、隐约融合的空间组合，一眼望不穿，一步走不完。设计师还巧妙地从古书籍的制作工艺汲取灵感，将活字印刷为切入点，将书架拆分为多个活动书格，利用可活动的折板，实现古书籍的展示、翻阅、保存的特殊要求。利用有限的空间进行重构。独个为字，组合成书，灵活组合，小到书盒，大到柜体，合理利用每一个空间。创造层次丰富，实用性强的书架形态，从而实现书店整体的现代中古朴的风格。

Design Concept

Drawing on the way of space division of Chinese classical gardens, designers change scenes by steps and see big things by small ones, which cleverly dissolve the uncomfortable feeling of plain and inflexible brought by the narrow and monotonous space. The space is divided into three spatial combinations with different virtual and real conditions, split, split and close, and vaguely integrated, which can not be worn at a glance and can not be finished in one step. The designer also cleverly drew inspiration from the production process of ancient books, took movable type printing as the entry point, split the shelf into multiple movable book cases, and made use of movable folding plates to realize the special requirements of display, browsing and preservation of ancient books. Use limited space to reconstruct. A single word, combined into a book, flexible combination, small to book box, large to cabinet, reasonable use of every space. Create a rich level, practical bookshelf form, so as to achieve the overall modern in the simple style of the bookstore.

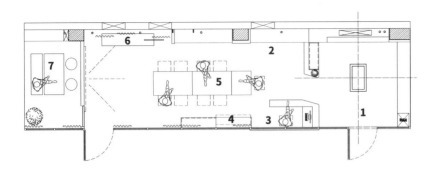

1 入口展示
2 书架陈列
3 收银台
4 展柜
5 活动区
6 展览区
7 VIP 接待

项目信息

项目名称：布衣古书局
地点：北京市东城区
南阳·共享际一层
委托方：布衣书局
设计单位：优享与行策划
咨询有限公司
设计师：王翔，吴雪
设计时间：2020 年 4 月～5 月
竣工时间：2020 年 7 月
项目面积：34 平方米

"塊"咖啡
项目介绍

"塊"咖啡是知名咖啡品牌 Cupone 孵化的全新冻干咖啡品牌，以"一块小宇宙"的品牌概念诠释新的品牌精神，设计完全匀质的白色空间，可以看作宇宙中完全匀质的黑色空间的反空间，与胡同青瓦灰墙的城市空间形成强烈对比，用抽象的空间、抽象的材料对比具象的立面、小尺寸的传统材料，强调"老"与"新"的对话。

抽象的材料：

尽量模糊材料的质感、尺寸。吊顶、地面选用具备大幅、哑光、无缝、超薄、耐磨、防水、防火特性的新型材料：白色岩板。配合完全匀质的照明：侧墙发光灯膜。

空间的对话：

中间的短墙将空间分为简明的"前"与"后"两个区域，对应"新式"与"老式"的咖啡制作，同时也将空间纵向分割成三份，分别提供给咖啡师、设备、顾客。

中国·北京　东城区东四北大街467号

Project Introduction

Famous coffee brand Cupone hatching new freeze-dried coffee brand, to the brand concept of "a small universe" new brand spirit, design completely homogeneous white space, can be viewed as completely homogeneous black space of space in the universe, in stark contrast to Hutong park kind of urban space, with abstract space, abstract materials contrast with concrete facades and traditional materials of small size, emphasizing the dialogue between "old" and "new".

Abstract material:

Fuzzy material texture, size as far as possible. Ceiling, ground selection with large, matte, seamless, ultra-thin, wear-resistant, waterproof, fire characteristics of the new material: white rock board. With fully uniform lighting: side wall light film.

Conversation in space:

The short wall in the middle divides the space into two simple "front" and "back" areas, corresponding to "new" and "old" coffee making, and also divides the space vertically into three parts for baristas, equipment and customers.

主理人说

林阳_一块小宇宙主理人

这不是一家传统意义上的咖啡厅，当你去到这里才可以感知到，没有太多可供歇脚的桌椅，里面是干干净净的全白色，有点科技感和未来感。我们实现了一块立体冻干，是为了让客人随时随处可以喝到高品质咖啡。这些做法看似很小众，但我们的愿望是服务大众。咖啡可以是这样的，恩，咖啡厅也可以是这样的！

项目名称："塊" 咖啡
地点：北京市东城区东四北大街467号
设计单位：空间站建筑师事务所
设计时间：2021年7月～8月
竣工时间：2021年10月
项目面积：39平方米

无瓦农场

项目背景

无瓦农场位于北京市昌平区小汤山镇尚信村汤尚路，毗邻温榆河森林公园河畔，地理位置优越，本身也具备丰富的农作物和植被资源。农场面积共 150 亩，一年四季都有丰富的果蔬在生长。目前，无瓦农场已从事生态农业 16 年之久，始终秉持着绝对自然的原则，种植出来的全部果蔬都是在自然的阳光雨露下培育而生，呈现的是最纯粹自然的状态，完全零农药、零化肥、零激素，保证绝对绿色健康。

中国·北京 昌平区尚信村

Project Background

WOW Glamping is located in Tangshang Road, Shangxin Village, Xiaotangshan Town, Changping District, adjacent to Wenyu River Forest Park riverside, the geographical position is superior, itself has rich crops and vegetation resources. The farm covers a total area of 150 mu, with abundant fruits and vegetables growing all the year round. At present, WOW Glamping has been engaged in ecological agriculture for 16 years, always adhering to the principle of absolute nature, all fruits and vegetables are cultivated under the natural sun and rain, presenting the most pure natural state, completely zero pesticides, zero chemical fertilizers, zero hormones, to ensure absolute green and healthy.

未来，无瓦农场将以"生机、生态、生活"三生融合的内涵为运营理念，通过其自身优秀的地理位置和自然条件，致力于打造以户外活动为基础，绿色农场景观作为背景，能够将户外拓展与科普教育相结合的创新型农业生态旅游基地。充分发挥昌平区的区域资源特色，在绿色消费领域树立标杆，打造精品乡村旅游思路，为乡村旅游及农业业态转型做出示范。通过寓教于乐的方式，提供生态教育的科普，为新时代消费者灌输潜移默化的生态环保意识。在实践中创新，在探索中成长。

Vision

In future, WOW Glamping will be "vitality, ecology, life," the connotation of the stroke of fusion for the business philosophy, through its excellent geographical location and natural conditions, is committed to build on the basis of outdoor activities, green farm landscape agriculture as a background, able to outdoor development combined with the popular science education base of innovative agricultural eco-tourism. Give full play to the regional resource characteristics of Changping District, set a benchmark in the field of green consumption, create high quality rural tourism ideas, and set a demonstration for rural tourism and agricultural business transformation. Through the way of edutainment, we provide ecological education and popular science, and imbue the new generation of consumers with imperceptible ecological awareness. Innovation in practice, growth in exploration.

运营者说

葛璐_共享际执行副总裁

在无瓦农场中，"学"无处不在，如农耕劳动体验、亲子小菜园等形式。因为从城市人需求的视角来看，久居城市的人们渴望了解农业的奥秘及农村的生活方式，为广大青少年提供了近距离接触农业、了解农业科技的绝佳场所，真正做到了寓教于乐。

钟宇辰_共享际CPO

对大自然的渴望，或者对自然界的无知，皆因缺乏时间到户外，特别是乡野田园所致，现实生活中，"自然缺失症"人群已经从儿童扩展到了成人。无瓦农场充分意识到自然教育对孩子乃至成人的素质教育重要性，无瓦不仅是京郊有机农业与自然教育的培育地，同时也将自身定位为渴望助力国家乡村振兴方针的实践者。从小事做起，以点带面，将中国传统的农业根基以现代创新的方式向外界推广。

设计理念

设计师利用无瓦农场现有的农耕用地、果林与少量的配套用房，将传统农业中的农耕场景、产品与现代休闲、田园生活方式巧妙结合，打造适合都市人度假的新农村消费场景。从农业种植与亲子教育为出发点，让农业更容易被体验，让农业更加休闲，让农事亲子教育更好地丰富家庭度假的选择，让农场不再仅仅依靠传统的耕种获得收获。

设计中更加注重人在农场与农事中的体验与学习，从人对行为与收获出发，挖掘农场种植内容，探索自然教育的多元创新，让更多的城市人获得健康果蔬的同时，亲近自然与农村，回归土地，收获知识与伙伴。

Design Concept

The designer makes use of the existing agricultural land, fruit forest and a small number of supporting houses to combine the farming scenes and products in traditional agriculture with modern leisure and pastoral lifestyle which create a new rural consumption scene suitable for urbanites to spend their holidays. Starting from agricultural planting and parent-child education, we will make agriculture easier to experience, make agriculture more leisure, enrich family vacation choices through farming education, and make farms no longer rely on traditional farming for harvest.

The design pays more attention to people's experience and learning in the farm and farming. Starting from people's behavior and harvest, it excavates the planting content of the farm and explores the diversified innovation of nature education. While enabling more urban people to get healthy fruits and vegetables, they get close to nature and countryside, return to the land and harvest knowledge and partners.

项目信息

项目名称：无瓦农场
地点：北京市昌平区尚信村
委托方：北京木兰景天生态科技有限公司
内容策划：优享与行策划咨询有限公司
设计单位：MOSS
设计师：Leng，张清发
设计时间：2021 年 3 月～4 月
竣工时间：2021 年 6 月
项目面积：6000 平方米

行业专家

随着城市更新成为国家战略，各行业的专业人士都将更加关注城市更新领域的发展与研究，作为成熟的城市更新和内容运营商，需要来自各个领域的专家、各级领导的指导与建议。

Experts

As urban renewal becomes a national strategy, professionals of all professions in the whole industry will pay more attention to the development and research of urban renewal field. As a mature urban renewal and content operator, guidance and suggestions from experts in various fields and leaders at all levels are needed.

单霁翔_中国文物学会会长，北京东城文化发展研究院院长

如果说建筑的外形、轮廓是城市风貌和城市文化的一部分，相对注重建筑和环境的关系，那么建筑的内部空间则体现了建筑的功能和建筑师的思想，更多反映了建筑和人的关系。在城市更新中，建筑内部空间的呈现可以为建筑和人们的生活带来更多的文化内涵。这是非常有意义的事业。

周　榕_清华大学副教授，中央美院客座教授，建筑与城市评论家

新时代的城市设计师如何变身为城市矩阵中合格的产品经理，如何理解时代、筹算人心、创意内容、匹配形式、运营社群、激活生态、玩转传播、夯实口碑……凡此种种综合性的谋略架构与企划成果，都可以在本书中找到丰富的实践案例。

吴　晨_北京市建筑设计研究院有限公司总建筑师

某种意义上，城市的文化复兴和几百年前西欧的文艺复兴有相似之处，城市的文化复兴与城市的商业繁荣并不冲突，反而应该是共生互促的关系。本书中所展示的很多建筑项目，为城市的文化复兴提供了极具洞察力的样本。

韩冬青_东南大学建筑学院教授，东南大学建筑设计研究院有限公司总建筑师

本书呈现的案例，十百千万，尺度不等；职住行消，各得其所；犹如镶嵌在既有城市肌理上的颗颗明珠，各耀其采。城市更新，起步维艰，方兴未艾。该书面世，恰逢其时！

廉毅锐_清华大学建筑设计院产业园区中心主任

更新，理论中有不同的名词和做法，但核心是，新的功能、新的人口激活原有地域真实复兴了场所。新功能顺利立足、蓬勃发展，在更新中是最为挑战之难题，甚至可以说是更新中一切之根本。

单　踊_东南大学建筑学院教授

共享际与优客工场定位是精准而恰当的，意识是清晰而敏锐的，方法也是务实而有效的。其思考与实践成果集成，为盘活城市的低效存量、助力城市的文化传扬，提供了不可多得的优质样本。

赵　辰_南京大学建筑与城规学院教授

从环境、城市、建筑的空间内容入手不断更新发展而策划、设计、运营，必将成为巨大社会需求的趋势，本书提供的正是在此领域具有先进意义的实践案例，值得阅读。

袁　牧_北京清华同衡规划设计研究院副院长，总规划师

这样一本来自使用者与运行者的城市更新著作弥足珍贵。城市的使用者和运维者在有机的城市更新中所承担的角色是任何高明的规划师建筑师都无法替代的，全社会共同参与的更新才会带来城市真正的复兴。

张　兵_弘石设计创始人，董事长

优客工场与共享际让房子不再是冰冷的盒子，给予建筑由里及表的非凡蜕变。在阅读时，无形中与一座座老建筑进行了一次次穿越时空的对话。让同样奋斗于城市更新一线的我们感到倍加振奋，也感受到了城市的美好未来。

王　辉_URBANUS都市实践建筑设计事务所创建合伙人

研发新的空间类型学，不只是让空间使用价值从被动的消费转型为主动的生产，更是要再生产出新型的社群关系和生产关系。因此，阅读这本书不要从产品或作品角度出发，而是要看到这些设计背后所担负的社会转型时期的使命。

丁沃沃_南京大学建筑与城市规划学院教授，博士生导师，院学术委员会主任委员

共享际与优客工场直面问题，实事求是，脚踏实地，以实际行动交出了一份高质量的答卷。当人们热议城市更新的时候，《城市更新与空间内容探索》一书不失为一份很好的教材。

刘军进_中国建筑科学研究院有限公司，建研科技股份有限公司总裁

翻阅完作品集，顿觉耳目一新，细细品来，体会有三：以人为本，运营为先，内容为王；创造趣味灵动空间，实现历史与现代共存；重视线上线下互动，流量和数据支撑运营。

王永平_商务部市场运行专家，全联房地产商会商业地产工作委员会会长

真正的城市更新不是喜新厌旧，而是城市文化的传承与发展，让老人看到新——新的建筑风貌与内容迭代，让新人看到老——老的历史文脉与文化积淀。世界各地值得称道的城市其魅力无不在于此。

翁　菱_综合艺术策划人，艺文中国联盟、艺文创新文化发展有限公司创始人

我相信社会的良性变迁需要有使命感、有创造力的人共同参与，共享际和优客工场便是其中之一。如今他们的工作成果无私地分享给大家，相信一定会带来启发和思考，成为社会发展的镜像切片。

冯斐菲_北京市城市规划设计研究院教授

曾经破旧的老建筑被巧妙地更新为吸睛的地标，曾经闲置的空间被赋予了最前沿和最时尚的内容。这就是共享际与优客工场！它又像一个个火种照亮和温暖着周边，无论在城市的街巷还是乡村的田野都能让人眼前一亮、心中一动！

陶红兵_北京愿景集团董事长

一张新画布上画最新最美的图画比较容易，让一张斑驳破败的旧画重新焕发光彩很难。城市更新所做的就是后者，这是在书斋或办公室里无法描绘指点出来的。

陈方勇_中国城市更新论坛秘书长，HuanXin创始合伙人

地产存量时代的到来让存量空间的改造成为主旋律，优客工场和共享际从创立之初就能够根据当下年轻人的工作、生活习惯创造出更具有现代感、开放融合、令人欣喜的创意空间，让富有活力的年轻人重新活跃在城市中心。

杨东辉_东南大学设计院风景园林院院长，东南大学建筑学院景观系副教授

优客工场与共享际把创新的运营理念引申到城市空间更新中，更加着眼于深层次的社区活力的营建，而不是简单的空间重构，不仅深度理解既有城市空间的文化延续、生活方式和老建筑具有的场景感，同时还通过创新性的内容注入，解决了城市更新过程中很难解决的自我孵化、持续经营以及未来的商业活力等一系列问题。

范小冲_阳光100集团副董事长

优客工场和共享际用自己的心得实践，图文并茂，为我们展现了中国城市更新的美好春天，吸引年轻人是城市更新的本质，也是本书的魅力所在，中国城市更新的大潮已来，未来更精彩。

郝　群_华贸集团控股有限公司总裁助理、商管总部副总裁

前瞻城市发展趋势与未来生活方式，汇集城市更新经典案例，探索面向新青年消费群体的新模式、新空间、新内容。

刘爱明_中城新产业控股集团有限公司董事长

书中呈现了很多精彩、有趣、有生命力的空间运营案例，仔细研读一定能体会到。

蒋　纹_浙江省建筑科学设计研究院有限公司总建筑师

对我国的城市更新而言，这本书就是一场"及时雨"。旧城能及时得到"内容"和"运营"的润泽，才能在"文化"和"经济"上获得再生。

王　永_品牌联盟董事长、中国品牌节创始人兼秘书长

每个城市的文化、个性不同，城市更新"更新什么""怎么更新"，除了顺应城市发展规律，同时还需要兼顾城市文化、个性中的生活"美"。本书将从空间与内容的独特角度，带着我们探索，为我们阐释城市文化、个性中的生活"美"。

张　刚_财视传媒CEO

他们在更新城市时，发掘和引领着新的生活方式。他们不是建筑师和简单意义的运营商，而是"城市更新师"的萌芽。正如梁思成所说，"中国的每一个城市，无论新旧，都必须计划和改善"。

隋秋实_Architecture Assistant at Foster and Partners

更新项目主要问题围绕在如何更新过去的建筑物以适应当前人们对效率、福祉和可持续性的新需求。本书见证了共享际与优客工场近些年对国内更新项目的探索历程，值得我们学习参考。

寄　语

城市，对于我们每一个人都是亲切又陌生的，城市承载着每一个人，却又是因为人的生活而有所不同，城市看起来是"城"与"市"的空间概念，为人提供防御的居所和交易的场所，但却在经历时间洗礼后，因人的归属与习性塑造了城市新的意义，它是文化，是历史，是细致入微的生活，是每一个生活在其中的活生生的人。

时间是一个悄然的塑造者，无声之中改变着人和城市。十年间的中国城市发生着翻天覆地的变化。空间的更新是有体有形的，但因其而改变的生活却是我们更新的本质所在。

我们希望创造一个培养皿，承载先锋内容，也创造卓越的生活方式！

Remarks

Urban is kind and strange for each one of us ,carrying every one in the city, but it still different because of people's lives, the urban seems to be the "protect" and "market" concept of space, which provide defense accommodation and trading places.But after the baptism of time, it shaped the city new meaning because of humans' belonging and habits. It is culture, is history.It's a nuanced life, it's every person living in it.

Time is a silent shaper, silently changing people and cities. In the past ten years, Chinese urban have undergone earth-shaking changes. The renewal of space is visible, but the life changed by it is the essence of our renewal.

We want to create a petri dish for vanward content, but also for a great lifestyle!

LIVE
WORK
PLAY

共享空间站式Co-Living共享生活社

为了追求更女
我们已经毁找
原已够好的

天地之间有许多事情
是你的睿智所无法触及的

真理是时间的孩子
不是权威的孩子